THE SOUTHWEST: A FIRE SURVEY

To the Last Smoke

SERIES BY STEPHEN J. PYNE

Volume 1, *Florida: A Fire Survey*

Volume 2, *California: A Fire Survey*

Volume 3, *The Northern Rockies: A Fire Survey*

Volume 4, *The Southwest: A Fire Survey*

STEPHEN J. PYNE

THE SOUTHWEST

A Fire Survey

TUCSON

The University of Arizona Press
www.uapress.arizona.edu

Printed in the United States of America
21 20 19 18 17 16 6 5 4 3 2 1

ISBN-13: 978-0-8165-3248-3 (paper)

Cover design by Leigh McDonald
Cover photo by Brady Smith: *Schultz Fire, 2010*. Courtesy of the USDA Forest Service, Coconino National Forest.

Library of Congress Cataloging-in-Publication Data
Names: Pyne, Stephen J., 1949– author. | Pyne, Stephen J., 1949– To the last smoke ; v. 4.
Title: The Southwest : a fire survey / Stephen J. Pyne.
Description: Tucson : The University of Arizona Press, 2016. | Series: To the last smoke / series by Stephen J. Pyne ; volume 4 | Includes bibliographical references and index.
Identifiers: LCCN 2016004471 | ISBN 9780816532483 (pbk. : alk. paper)
Subjects: LCSH: Wildfires—Arizona—History. | Wildfires—Southwest, New—History. | Wildfires—Arizona—Prevention and control—History. | Wildfires—Southwest, New—Prevention and control—History. | Forest fires—Arizona—History. | Forest fires—Southwest, New—History. | Forest fires—Arizona—Prevention and control—History. | Forest fires—Southwest, New—Prevention and control—History.
Classification: LCC SD421.32.A6 P96 2016 | DDC 363.370979—dc23 LC record available at http://lccn.loc.gov/2016004471

♾ This paper meets the requirements of ANSI/NISO Z39.48-1992 (Permanence of Paper).

To Sonja,
old flame, eternal flame

CONTENTS

SERIES PREFACE

To the Last Smoke

WHEN I DETERMINED to write the fire history of America in recent times, I conceived the project in two voices. One was the narrative voice of a play-by-play announcer. *Between Two Fires: A Fire History of Contemporary America* would relate what happened, when, where, and to and by whom. Because of its scope it pivoted around ideas and institutions, and its major characters were fires or fire seasons. It viewed the American fire scene from the perspective of a surveillance satellite.

The other voice was that of a color commentator. I called it *To the Last Smoke*, and it would poke around in the pixels and polygons of particular practices, places, and persons. My original belief was that it would assume the form of an anthology of essays and would match the narrative play-by-play in bulk. But that didn't happen. Instead the essays proliferated and began to self-organize into regions.

I began with the major hearths of American fire, where a fire culture gave a distinctive hue to fire practices. That pointed to Florida, California, and the Northern Rockies, and to that oft-overlooked hearth around the Flint Hills of the Great Plains. I added the Southwest because that was the region I thought I knew best. But there were stray essays that needed to be corralled into a volume, and there were all those relevant regions that needed at least token treatment. Some like the Lake States and Northeast no longer commanded the national scene as they once had, but their stories were interesting and needed recording, or like the

Pacific Northwest or the oak woodlands spoke to the evolution of fire's American century in a distinctive way. I would include as many as possible into a grand suite of short books.

Each regional survey would have an organizing frame and some internal logic to its choice and arrangement of essays. As the project matured, I began to see it as a transect through an era of American pyrogeography, a kind of literary CAT scan, a national reconnaissance told through a single voice and vision that could serve as a baseline for future fire folks. Unique, maybe quirky, the project pretends to be nothing more than a sampler and cross section; but nothing like it has been done before. I know that I would have welcomed such a series as I began my own research into the American way of fire.

My original title now referred to that suite, not to a single volume, but I kept it because it seemed appropriate and because it resonated with my own relationship to fire. I began my career as a smokechaser on the North Rim of the Grand Canyon in 1967. That was the last year the National Park Service hewed to the 10 a.m. policy and we rookies were enjoined to stay with every fire until "the last smoke" was out. By the time the series appears, 50 years will have passed since that inaugural summer. I no longer fight fire; I long ago traded in my pulaski for a pencil. But I have continued to engage it with mind and heart, and this unique survey of regional pyrogeography is my way of staying with it to the end.

Funding for the project came from the U.S. Forest Service, the Department of the Interior, and the Joint Fire Science Program. I'm grateful to them all for their support. And of course the University of Arizona Press deserves praise as well as thanks for seeing the resulting texts into print.

PREFACE TO VOLUME 4

MY SOUTHWEST SURVEY has been the most fragmented of the regional studies to write. I live here. I thought I knew the basics, could jog to various sites as opportunities and weekends allowed, could find a home for a few orphaned essays, could put the suite of essays together without the concentrated travel that has characterized my research elsewhere. I did it, but I was wrong.

I also assumed the Southwest would be published after the Big Four regions. It would have been more efficient and more thematically coherent to have organized a targeted trip as I did for the other regions. Instead, I have a more fractured collection written across several years, though mostly between 2011 and 2014, and published out of sequence to the order in which they were written. Several pieces were written as stand-alone essays on fire themes, not as part of a regional reconnaissance. While they have been rewritten, retrofits are never as elegant as pieces designed for a collective purpose from their origins.

Yes, a tiles-for-a-mosaic quality fits a major theme of Southwest fire, and, yes, the splintered chronology also aligns with the Southwest's lumpy history, but it all comes perilously close to the imitative fallacy. I know only too well how much got left out. But the result is a good sampler of the southwestern fire scene. I'm pleased I had the chance to write it.

THE SOUTHWEST: A FIRE SURVEY

Map of the Southwest.

PROLOGUE

Cycles of Fire

THE SOUTHWEST IS BUILT TO BURN. Pick any facet of any fire triangle you wish and you will find it here, often accented. This is a place where the elements meet but don't merge.[1]

Air. The region undergoes an annual cadence of wet winter and wet summer with a long rainless spring between. Overlying this are the decadal rhythms of drought and deluge typical of El Niño's comings and goings. Still wider waves of deeper drought appear and fade, some notorious as century-blasting megadroughts that can drain settlements as they do biotas. All fire weather lacks is an eruptive regional wind like the Santa Ana or mistral to drive flame through landscapes. Instead it relies on powerful spring southwesterlies and downdrafts from early season thunderstorms that stir up dust storms over deserts and fire storms over flames.

Earth. The Four Corners beloved of political cartographers might equally stand for four great geomorphic provinces that converge on the region. From the northeast, the Rockies. From the east, the Great Plains. From the north, the Colorado Plateau. From the west, the Basin Range. Save the plains, it's a landscape of abrupt landforms, difficult to access, not prone to slick segues or gradual blurrings. Outside the plains the New Mexican horizon is visually granular, rich with discontinuities, as landscape tile abuts tile, often with little transition. The horizon lifts and sinks, a sudden emptying and filling of the scene with plain and peak, gorge and mesa. Every twist in the road exposes a new vista. If the region's famously clear air allows you to see long distances, it also sharpens the borders of what is seen. It's a place of abrupt edges: one landform ends and another begins, the biota splashes in patches and layers up mountains like a ziggurat.

The angular jumble of mountains, valleys, and plateaus means that there is always someplace dry and someplace where lightning is temporarily segregated from rain. It's a complex geography of fire, full of niches, of quirky topography and odd pulses amid broad regional rhythms, all of which add up to an abundance of fires, mostly small, but occasionally large, most brief but some that can creep and sweep over months and across mountains. If the winter is wet, the lowlands are flush with growth, which becomes fine fuels in the spring. If the winter is dry, the highland forests are leached of moisture and primed to burn. Year in and year out, something is always available to burn, and something nearly always does.

Water. As if emulating its earth, the Southwest's water comes and goes abruptly. The summer storms can drench one side of a street and leave the other dry. One season can drip with moisture, the next lack a drop. When water gathers, it flows in well-delineated channels with scant wetlands or floodplains. Floods come in flashes down dry arroyos, not with slow crestings and fallings. Reservoirs place water directly against rock. In the eastern United States lakes can seem a thickening of water gathered in the ground and through biotas. In the Southwest there is no intermediary: water abuts stone and air.

Mostly it's a place of aridity. A biota has evolved to suit this complex, oft-sudden rise and fall of land and of rain. Growth comes in some seasons, decomposition in others. The dead biomass can build up, locking critical nutrients like phosphorus into inert chunks, slowly pulverized by physical impacts. To recycle those scarce nutrients the ecosystems rely on fire. Save the most rocky and isolated sites, everything will sooner or later burn. The Sonoran Desert might know fire three to four times a century; the wide-sweeping pineries of the Mogollon Rim, perhaps every three to six years. Some of the high grasslands and ponderosa savannas probably burned as often as fuels existed to carry the flames.

With water, as with air and earth, transitions are sudden. The rhythm of the rains seems to toggle on and off. One year they shower and grow combustibles, the next rains retire and allow those hydrocarbons to burn. A series of wet years followed by a sudden drought is guaranteed to spark a region-wide epidemic of fires.

Fire. Its fires are as much a part of the Southwest matrix as its mesas and monsoons. Its terraced rims rising from the desert to the Colorado Plateau have the heaviest concentration of lightning fire in the nation, rivaled only by central Florida. But since the modern climate began to congeal, and the monsoon established itself, around 9,000 years ago, people have been on the scene and have added their fires to nature's. Between them the region enjoys nearly constant ignition, awaiting only the right alignment of fuels and winds to burn.[2]

People used fire in hearths to cook, heat, light, to work wood and stone, to produce smoke to ward off insects. Here and there, by calculation and carelessness, they burned niches and sometimes panoramic landscapes to help hunt, to freshen spring fodder, and to assist with foraging. They burned for wood rats along Colorado River tules and for deer amid pine steppes. In this enterprise, they had biotic allies; the extinction of megafauna encouraged more browse and pasture for fire, which further leveraged the power of the torch. They burned pinyon-dominated landscapes to help harvest pine nuts. They burned small plots for gardens. They used fire to signal, fire for ambush and war, fire for ceremony. And they littered the landscapes they inhabited or traversed with campfires and the odd spark. Surveying Arizona's forest reserves in 1902, S. J. Holsinger declared that "the most potent and powerful weapon in the hands of these aborigines was the firebrand." They kept a land ever ready to burn ever simmering with fire.[3]

The upshot was, historically, a fire-drenched scene and a fire history that flickered and flared across millennia. But that ancien régime shattered in the 1880s with the revolutionary arrival of steam. An unnatural quiet followed; fires quelled as the grasses that fed them went into the gullets of sheep and cattle. Then a new regime rose out of the overgrown ruins. By the 1970s, even after grazing had decimated the historic fire regimes and the Forest Service had installed aggressive fire suppression, the Southwest's national forests averaged more fires per year than any other region, had the second-highest rate of burned acreage, and experienced critical fire weather with greater frequency and persistence than anywhere else in the United States. As the pressures have eased—as grazing recedes as a forcing mechanism, as restoration replaces suppression, as climatic cycles deepen and crowd together—fire has returned in force. The ancient pattern of surface burning, dappled with many small fires or long-lingering burns, is yielding to more massive, high-intensity complexes. By the 1990s that transition had begun; and in keeping with the Southwest's character, the switch came suddenly.[4]

The Southwest is a landscape of distinct parts. Even when pieces merge into larger wholes, those wholes function as a piece that can seem remarkably separate and distinct. For the land, the sky island is the perfect embodiment of this phenomenon. For people, it is the reservation, whether legally gazetted or a de facto rooting. For fire management the public domain parceled into national forests, tribal reserves, wildlife refuges, state trust lands, military posts and proving grounds, and checkerboard Bureau of Land Management (BLM) holdings.

Every part of the Southwest fire scene can be found elsewhere. The prairies burn more often. The Great Basin and parts of California display an equally striking (and even more exaggerated) abruptness in the transitions between landforms. Florida abounds in lightning. What makes the Southwest distinctive is not only how the pieces jumble together but how they persist. It retains the relics of its past—the fabled aridity that encourages fire also retards decay. Nowhere else is the record of humans, or of fire, so well preserved. The Southwest's fire cycle is a cadence of history, a palimpsest of cycle upon cycle.

The character of its countryside is the template for its human settlement, which is ancient, rudely rhythmic, and preserved. The first hard relics, from Clovis peoples, date to 13,000 BP. (The modern monsoon climate shuffled into existence around 9,000 years BP. The ponderosa pine forest dates to roughly 6,000 years BP.) In the southeast and Mississippi Valley generally, the record of the mound builders sank into the soil. In the Southwest Anasazi ruins fill cliffs and tattoo mesas; the ancient canals of the Hohokam can still be traced; terraces a millennium old contour hillsides. Oraibi is the oldest continually inhabited town in the United States; the pueblos of the upper Rio Grande, the longest occupied places by Hispanics. Lithics lie promiscuously on the ground. Even adobe endures. Log corrals stand, as though chiseled from rock. Where fires have burned old habitations, the charcoal remains and, an added bonus, allows for dating.

But persistence characterizes the people not just their artifacts. Each newcomer, each wave of settlement, piled on the old without completely erasing it. Through cycle after cycle of conquest, the indigenes endured. They changed aspects of their culture, yet never assimilated completely, and by claiming homelands they retained their identity. Its peoples like its landscapes form a kind of caste system, each present but not melted and forged into a new state. There is no cohesive middle: the region is a mosaic. The indigenes survive in reservations. Because of them the Southwest boasts a living record of human history.

The same holds for its fires: not just charcoal but the wood with it tends to endure. In rare cases the wood has turned to stone (some of the logs in the Petrified Forest even show fire scars). Fire scarring in ponderosa pines, particularly, track fire history back centuries. But that history can be pushed further into the past by examining still-extant stumps or, more astonishingly, the vigas and pillars amid ruins. By correlating living tree records with patches of the past left in preserved wood, the annals can reach back almost 1,500 years. Only the giant sequoias hold a longer continuous chronicle. The Southwest even boasts an unexpected climatic record through packrat middens, essentially mummified by aridity.[5]

So arise the two thematic axes of Southwest fire history. The cycles of burning echo the cycles of history. People and fire—the old regimes persist within and because of reserves.

It's one thing to have lots of fires, another to influence fire management throughout the country. What happened nationally affected the region because it was refracted through federal institutions, and so, too, what happened in the region could influence national thinking because those same institutions could broadcast across the country. Because the old did not dissolve, because those chunks were reserved and preserved, the Southwest was politically difficult to digest, but because Arizona and New Mexico were not admitted to the Union until 1912, the last of the continental states to join, state-sponsored conservation was already in place. Federal forests and parks didn't have to cope with landscapes already speckled with private holdings, save along the Rio Grande, or wait for abandoned lands after clear-cuts or following economic collapse during the Depression. Teddy Roosevelt doubled the size of the national forest system before he left office; his Governors Conference on Conservation occurred four years before Arizona gained admission. The

Southwest boasted the first experimental range, the Santa Rita, in 1903; the first experimental forest, at Fort Valley, in 1908; the first laboratory for dendrochronology, at the University of Arizona, in 1937. Compared with other regions the Southwest had a federal presence that predated Anglo settlement, retains lots of federal land, and holds that land in unusually coherent patches. Compare its evolution to west Texas to see what national integration, when it came, could mean. However anarchic its many public domains may appear it has massed ownership in ways impossible in the privately owned dominions of the eastern United States.[6]

Fire exclusion had begun with overgrazing, which commenced when the last Apaches were rounded up. But firefighting as an organized activity was a charge from the onset. The Southwest's remoteness and sparse settlement invited early-adoption mechanization. Radios were used on firelines in the Apache National Forest and automobiles patrolled the Coronado in 1916; airplanes flew reconnaissance over the Catalinas in 1921; bulldozers shouldered firelines and roads through hillsides in 1928, little more than a year after they were invented. By then the Southwest had become a miniature for the self-destructing Old West: its grasses eaten out, its springs dried up, its game gone, its indigenes diminished and sequestered, its fires lost.

Yet fire protection seemed to stand outside other criticisms. The great prophet of 20th-century environmentalism, Aldo Leopold, began his rangering career in the Southwest, from which experiences he drew some of his most celebrated essays. On the Apache he recorded the killing of the last grizzly (on Mount Escudilla), and on the East Fork of the Black River he shot the wolf he made famous in "Thinking Like a Mountain," while in New Mexico he recorded the horrific scars—geologic in their savagery—wrought by overgrazing. He pushed hard to formally gazette the Gila Primitive Area as the nation's first wilderness.

But that did not mean he supported fire. The region experimented with let-burning and rejected the idea, and Leopold himself denounced the notion of light-burning. It would, he insisted, undo all that the Forest Service had sought to reform by bringing forestry and conservation to the land. Years after he shot the she-wolf, he brilliantly reconsidered the role of predators; and he made the Kaibab deer irruption an international symbol of the upheaval predator control could wreak. But he never made a comparable transition for fire. He might come to tolerate it, but he did not promote it. In 1950 the Southwest instead gave us Little Smokey.

From time to time the region continued to experiment with controlled fire. After Harold Weaver went to Fort Apache in 1948 and two years later became regional forester for the Bureau of Indian Affairs (BIA), he inspired a program of broadcast burning in ponderosa pine, and for almost 30 years Fort Apache Reservation under Harry Kallander was the wunderkind of prescribed fire in the West. The experiments never left the reservation, however, and by the 1970s, after Kallander had retired, the project was fading under a kind of bureaucratic wasting disease. Mostly, when agencies undertook burning they hoped to support commodity production. Fire might sweep brush out of the Verde Basin and restore grass for herders; it could dampen the slash left by industrial logging in the postwar era. But not until the Gila adapted federal policies to allow confining and containing wildfires and made the Mogollon Mountains a model for using wildland fire to restore fire to wildlands did the modern era come to roost.

By then the Southwest was back in the national fire scene. The Jemez Plateau was morphing into a poster child for wildfire as a transformative feature; the Mogollon Rim, for unhealthy forests as a consequence of fire exclusion. The 2000 escaped burn that became the Cerro Grande disaster checked enthusiasm for prescribed fire. Megafires began to tumble over one another; every couple of years the size of the largest fire on record doubled. For New Mexico the 2000 Cerro Grande fire set the bar at 48,000 acres. The 2011 Las Conchas fire tripled that area (150,000 acres). The 2012 Whitewater-Baldy burn doubled it again to 298,000 acres. For Arizona, the 1996 Lone fire established the record at 60,000 acres until the 2004 Willow fire doubled it (120,000 acres), and the 2005 Cave Creek complex doubled that new high (244,000), and the 2011 Wallow fire doubled it again (538,000). (The 2002 Rodeo-Chediski fire was two separate fires with enough area between them—land deemed too unsafe to try fire suppression—to constitute a third.) Even in a country with a potential billion burnable acres, those kinds of numbers get attention.

The region responded with nationally noted efforts at remediation. Northern Arizona became a center for inquiries into forest health and the quest for possible remedies; uniquely, the so-called Flagstaff model pioneered techniques not only in silviculture but in the politics of social expectations. The Four Forests Restoration Initiative is the largest experiment conducted under the auspices of the Collaborative Forest Landscapes Restoration Program. In southern Arizona FireScape established a paradigm for new styles of fire restoration. The Gila continued to break ground in fire management as it pushed beyond montane woodlands into mixed-conifer highlands, tacked against climate change, and shouldered the burden of legacy landscapes.[7]

But more than climate change resisted. Tentacles from the region's metropolises—Phoenix, Tucson, Albuquerque—wriggled into mountains, high deserts, and scrublands, and what they didn't remake of old rural hamlets they made afresh. Against them rose a tribe of environmental activists who added sprawl to their old nemeses, logging and herding. The Center for Biological Diversity carried the timber wars of the Northwest to the Southwest. Indigenous peoples, too, joined the cause, speaking in their local tongues: national clashes broke out at Blue Lake between Taos Pueblo and the Forest Service, the San Francisco Peaks between Navajos and the Snow Bowl ski resort, and Mount Graham between Apaches who wanted the peaks left alone and the University of Arizona, which was determined to cap them with telescopes. Environmental legislation and the American Indian Religious Freedom Act gave political heft to the protest.

But fire didn't listen—didn't care. It responded to the conditions that existed regardless of cause or unintended consequence. Between 1994 and 2011 some 20 percent of Arizona's forests were overrun by high-intensity megafires; the percentage was higher in New Mexico; and these were states with scant forest to spare and those forests vital to urban watersheds. There was little point to creating biological reserves if they were blasted by fires well outside historic or evolutionary ranges. The bad news, however, was that the region needed much more fire, and would get it willed or not, but seemed likely to get it as bad fire. Less and less could critics use fire to animate some other message because, for everyone, fire itself was becoming the message that trumped all others. The thin reed of hope was that, if the various partisans could not rally around the need to promote good fire, they might find common cause in the threat posed by bad fire.

But good or bad, the Southwest would burn. And every change people had introduced over the past century, whether by choice or chance, whether from altering the climate or stuffing the woods with second homes, had pushed the scene toward fires that were more savage, less useful, and more environmentally ruinous.

SACRED MOUNTAINS

THE SOUTHWEST IS A JUMBLE OF TERRAINS—plains and plateaus, ravines and rims, gorges and peaks. The best-known features, however, are its canyons and its mountains, and these have been sacred to the various peoples who have lived in the region. The Hopis place the *sipapu*, the orifice in the earth from which humans emerged, in the gorge of the Little Colorado River. The Tohono O'odham claimed Baboquivari for a sacred mountain. The Apaches had Mount Graham. The Navajos bounded their ancestral lands by four mountains: Mount Blanca (*Tsisnaajini*) to the east; Mount Taylor (*Tsoodzil*) to the south; the San Francisco Peaks (*Doko'oosliid*) to the west; and Mount Hesperus (*Dibe nitsia*) to the north.

Americans have their own version. The Colorado Plateau has a greater density of national parks and monuments than any other geomorphic region. The Grand Canyon is part of a pantheon of venerated places. And the wildland fire community, too, might be said to have its own sacred mountains that bound the cardinal corners of how they imagine their landscapes of fire and define their place within them.

My candidates are the Jemez Mountains and the Mogollon Mountains of New Mexico, and the Huachuca Mountains and Kaibab Plateau of Arizona.

THE JEMEZ

Genesis Effect

STAND ON THE RIM of Cochití Canyon and look over a blasted landscape. What once was a densely forested gorge, a sky island on its head, is now stripped clean, save for ghostly legions of trunks. Moonscape is a common journalistic trope to describe postburn scenes; but those landscapes return, often with more ecological vigor than before. Here even the ash is gone; and only in niches in Cochití can you find a patch of green. There are no squirrels or rabbits scurrying among the ruins. There are no bark beetles amid the boles. There are no birds; no juncos, no red-shafted flickers, and because there is no carrion no vultures soaring on thermals. Even the last-survivor ravens have fled. There is silence ruffled here and there by wind. There is only rock, and the trunks like lithified pillars, and occasional pulses of floodwaters, the geologic matrix out of which an ecosystem might sometime return.[1]

It's a setting in search of meaning, or at least a metaphor. To natural scientists, it's a land being reorganized. To novelists, it's an apocalyptic, nuked landscape. To those who like their science and fiction fused, it's as though the Genesis device, featured in Star Trek's *The Wrath of Khan*, had unleashed its fiery front and destroyed the preexisting world in "favor of its new matrix."

It's a place to contemplate the future. Are the serial fires that are blasting the Jemez Mountains, and where the fires return, compounding one burn with another, a new normal? Or are they a biotic outlier as peculiar as the Jemez is a geologic one? Is the Jemez a sacrifice zone of the Anthropocene? Or the first tremors of a global Götterdämmerung that signifies the end of the world as we know it?

The Jemez Mountains are one of two vestigial megavolcanoes in America. In broad terms they mark a spot where the Rockies and the Colorado Plateau meet atop buried rifts before spilling into the Basin Range to the south. Over the course of 14 million years the Jemez has spilled lava and belched tuff and gases and built up an immense pile in northern New Mexico. It remained active during the Quaternary. There were major eruptions 1.6 and 1.25 million years ago before, roughly a million years ago, the mountain blasted lava, ash, and pyroclastics on such a scale that it blew away its own top and the great mass collapsed into a caldera. A resurgent dome rose at the center, the crater filled with flowing rhyolite, and later bursts surrounded most of the central massif with a daisy chain of smaller domes, leaving a landscape of grassy plains and forested hills known as the Valles Caldera.[2]

Since then the Jemez Mountains have spilled biological rather than geologic fire. From the onset of the modern monsoon climate, the landscape has been saturated with ignition. When wet-dry cycles stoke suitable fuels and winds whip and splash over the great conifer-coated massif, fires break out and slosh around the *valles*, over domes, across flanking mesas, and down gorges. The older record is chronicled in soil and lake charcoal; the newer, over the past centuries, in fire-scarred ponderosa pine; the most recent, in texts and photographs.

For millennia burn had passed over burn, leaving strata of ecological history. In some places the biota thickened, the equivalent of rhyolite domes. Some years had almost no spreading fires. A few, as in 1748, had many, or fires that crept and swept across the whole of the range, dodging rains and grasping for winds, over the long summer months. Some 23 times back-to-back years managed to scorch virtually everything. But throughout that meticulous annal, the forests and grasses and patches of scrub persisted. Through most of the mountains the fires scrubbed the biotic grime of 5 to 10 years from the surface. Mixed-conifer woods trimmed the crest of the highest rims, and when they burned, the flames wiped out stands, but such bursts were little more than vignettes decorating the margins of the text. What the fires removed either survived or was effectively replaced. The material record testifies that fires, and fire years, swelled and contracted, but always within bounds. The biota persisted through layer after layer of surface fire flows.

The Jemez Mountains are among the most intensively inventoried landscapes in America. Few places have a fire history as thoroughly documented: there is little dispute about the character of the fires that populated the Holocene. Century by century, millennium after millennium, the cadence of the fires continued, like hot springs fed by a deep chamber of climate.

Then that rhythm broke. In the late 19th century an irruption of livestock stripped out the grasses and stopped fire stone cold. After 1883 the annal effectively ends. Afterwards, although ignitions might sprout like lupines, they failed to spread. Their fuel went into sheep and cattle, while flames starved and expired. The record of fires simply ends, as though the recording needle ran out of ink. In 1905 the U.S. Forest Service assumed control over the forest reserves, and instigated a program of systematic fire exclusion. People could no longer set fires freely, and fires of any source that did appear would be suppressed. Though the core of the Jemez, most of the Valles Caldera, was held privately under the Baca Grant and its successors, its fires beat to the new regime.

The rhythm broke a second time a century later when, beginning in 1977, savage, woods-stripping, with a severity unprecedented over the duration of the Holocene, fires began to blotch the mesas and domes. When such fires reburned the same site, they vaporized wood and shrub and grass, baking the caldera like a ceramic bowl. Such serial fires no longer inscribed a living record on the land but obliterated the record altogether. The Jemez began shuffling from one of the best-documented landscapes in America into one that threatened to become a tabula rasa. History wasn't stopped, or even reversed. At places like Cochití Canyon, it was extinguished.

Behind this transformation lay new sources of power. The plateau had long been subject to seasonal transhumance of local flocks, though their size, and hence their impact, had been limited by the capacity of the local communities to absorb them. The market was local and, compared to the landscape, small. Shepherds burned, but along the transhumant corridors of seasonal travel and within established fields of fire. The flocks competed with fires, but there was ample room for both.

Then in 1878 the Atchison and Topeka railroad crested over Raton Pass. A railway reached Albuquerque in 1880, while a spur line brought locomotives puffing into Santa Fe. The next year a complete east-west connection across the continent was possible using those routes. In 1883 the Jemez experienced its last widespread fire season. (In 1885 the Atlantic and Pacific Railroad completed a more direct route.) Those rail lines proved a chasm in combustion history as profound as the rifts underlying the massif.

The ancient isolation of northern New Mexico ended. Flocks and herds no longer had to accommodate into the tidy string of pueblos along the headwaters of the Rio Grande: they had the nation for a market. They expanded to meet that demand. The problem was that the fuels feeding the flocks was far more finite than the fuels stoking the locomotives. While the coming of the U.S. Forest Service in 1905 changed the patterns of indigenous burning, it did nothing to alter the onset of the new combustion regime which had its sources and sinks elsewhere. Industrial combustion had a magma chamber thousands of times larger than that beneath the Valles Caldera. Over rails and roads it spilled out surface flows of internal combustion, while wiping the land clean of open flame.

The larger landscape began to adjust to the absence of routine fire. An ancient predator had been removed, and the population of conifers that it had held in check now exploded exponentially. The most dramatic changes occurred in the montane belt of ponderosa pine. Land that had held trees per acre that could be counted on a ranger's fingers and toes now numbered in the hundreds, and in places, the thousands. Locally, as around the forested caldera, abusive logging shattered the structure of the forest and left behind a scrub of conifer thickets. Combustibles stockpiled, now so congested that a flame in any particle could ignite a score of others, and they other scores in what might aptly be described as a chain reaction of combustion. Suppression could shackle the dragon only through good luck and a favorable climate.

In the 1960s a revolution in thinking, and then in policy, sought to reverse fire's exclusion with its controlled reintroduction. That protest flared most spectacularly in Florida and California; the Southwest contributed little directly; and here suppression remained the rule. By the 1970s the deteriorating scene inspired regional discussions, and by the 1980s some experimentation. Most of it occurred in the Mogollons, not the Jemez. If viewed from space, the Jemez would seem ideal for fire restoration: a massive island that if not wholly self-contained was an environmentally bounded unit. The reality was different.

Whatever the sentiment of reformist fire officers and creative ecologists, the freedom to maneuver at the Jemez was less than it appeared from afar. There were complexities hidden among the pixels and polygons. There were ruins on the surface, modern pueblos around the perimeter (the wildland-urban interface [WUI] of the Jemez), and a deep distortion in regional pyrogeography known as the Los Alamos National Laboratory (LANL). The lab had pioneered, with the atomic bomb, the explosive invention of an alternative energy to combustion. In fact, much of the damage from weaponized nukes came from the fires that followed, and nuclear power never displaced combustion when tamed into reactors (or even came close). But the lab has acted in the space-time of the Jemez fire field like a white dwarf star, deforming everything around it. Among federal facilities probably only Merritt Island, which houses the Kennedy Space Center within a fire-prone wildlife refuge, approximates the peculiar dynamics of the Jemez.

With almost malevolent cunning the lab and the town that serviced it were sited for maximum risk. They rose amid a forest unburned since the 1880s, and then kept unburned, save for a patch here and there every 20 years or so. It sat on the eastern flank of the massif, where it could receive fires driven by the prevailing spring winds from the southwest. And its secretive nature and lethal pollutants made treatments unlikely and prescribed burning unwanted. It resided amid a natural dump of combustibles that could burn with the energy output of Little Boy and Fat Man combined, though released over hours rather than nanoseconds. It made political sense to house the project in a remote, secure location. It made no ecological sense. The lab sat amid nature's more benign version of

U-235. The mechanical muscle of wildland fire protection, still bulked up with postwar military hardware, shielded the lab and its anomalous town from the consequences of the hasty decision to staff the Manhattan Project on the slopes of a combustible biovolcano. The lab was created by a wartime emergency. It was protected by a Cold War fervor for fire control.

Then on June 16, 1977, lightning kindled a fire on the eponymously named Burnt Mesa in the uplands of Bandelier National Monument. It escaped control and, with prevailing southwesterly winds behind it, fresh with spring vigor, it barreled over mesas laden with Anasazi ruins and a century of weapons-grade biofuels and bolted toward the Los Alamos National Laboratory.

The La Mesa fire announced a new era in the pyric history of the Jemez. The Park Service repurposed archaeologists as line scouts to keep bulldozers from flattening the ruins that dappled the mesas, insisting that there was no point in destroying the artifacts that were the reason for the monument's existence—more subtly, perhaps, challenging the unquestioned supremacy of national security needs over all other goals. A symposium followed which put down markers for the monitoring of future change. And the fire, though halted before it reached LANL, effectively posted a notice next to the others along the lab's border. The Department of Energy signs warned against intruders. The La Mesa signs, in the form of blackened snags, dared the fence to keep fire out.[3]

The fires paused as the region went into a wet cycle. It was a grace period, but while Bandelier prepared plans and began to light prescribed fires to counter the vast imbalances accrued over the past century, the Santa Fe National Forest did not, and LANL appeared to dismiss the La Mesa outbreak as an outlier, as though it were the outcome of an experimental error. The wet cycle dried up. In 1996 a dry cycle returned with pent-up ferocity. The Dome fire, boiling out of an abandoned campfire on April 26, again roared toward Los Alamos. A spring fire, it burned hotter and faster. This time a thin red line of backfires held it on Obsidian Ridge within shouting range of Los Alamos, only because the winds had providentially spun to blow from the north. Had the prevailing southwesterlies held, the fire would have ripped through lab and town.

The numbers were starting to add up, as though the Jemez were venting a lost century of fires in explosive spasms. The La Mesa fire had burned 15,444 acres; the Dome, 16,516; but far larger landscapes, laden with aching fuels, lay unburned. More than that it was obvious that the Pajarito Plateau was in the firing line: it lay in a seasonal wind tunnel primed with explosive matter that only needed a spark, an ignition device, to set it off. Officials were looking at regime change. Two years later fire officers and lab officials met to discuss some perimeter barriers in the form of fuelbreaks. The treatments were still underway when a prescribed fire near Cerro Grande in the uplands of Bandelier National Monument escaped control on the evening of May 4, 2000.

The suppression strategy elected to back off to Highway 4 and burn out. That burnout was lost, and the fire bolted out of the monument altogether and soared over the Pajarito Plateau toward Los Alamos. A fumbled suppression strategy again failed to halt the flames or to actively protect the town. One head fire crossed untreated borders to blow into the lab; another crept through the western half of the town incinerating houses by the hundreds. The fire reached 47,000 acres—the largest forest fire on record for New Mexico. Final costs set national records thanks to aggressive suppression and the Cerro Grande Fire Assistance Act; most of it went to rebuild the town. Blame scorched the air, but there was no escaping the fact that the prime mover was the botched prescribed burn.

The worst, however, was yet to come. On June 26, 2011, a dead aspen fell across a power line and kindled a fire amid drought, wind, and more than a century of accumulated combustibles. The plume twisted, held by horizontal roll vortices, as it rumbled across the caldera like a slow-churning tornado on its side. When the vortices faltered, a mushroom-cloud plume rose, and when the plume collapsed, as though it were a pyric thunderhead, it sent the ash, gases, and flames downward in a rush, a biotic flashover. The effect resembled a pyroclastic outflow that stripped everything. Where that rush channeled into gorges like Cochití, it poured through like a flood of pyric debris. Where it splashed over mesas, even those stocked with dispersed junipers little prone to burn, it immolated the landscape. Where it passed over lands fried by the Dome fire, it took everything down to rock. The Dome fire had stripped out the forests. The Las Conchas fire wiped away the black locust and Gambel oak and exotic grasses that had grown in its wake. By the time the rush ended, Las Conchas had tripled the size of the Cerro Grande fire.

The latent horror was that such a fire would blast through the lab and liberate stored or buried materials including plutonium that would contaminate the land for decades and, worse, poison the municipal watersheds of downstream communities. But even conventional incendiaries were sufficient to render the Rio Grande unfit as a source of potable water for 20 days at Santa Fe and 40 at Albuquerque. Nature's nemesis had answered human hubris with a force majeure against which resistance was futile. It seemed as though the Jemez had again found new avatars of eruptive energy.

That ignored the deeper driver. Las Conchas had started when the old source of natural power, free-burning fire, had crossed the new one, industrial combustion. An older order had fallen on a newer one. The 115 kilovolt transmission line that sent the spark over the fallen aspen ran power for the Jemez Mountains Electric Cooperative, which received its energy from coal-fired dynamos throughout the greater region.

With its dramatic prominence amid the northern New Mexico landscape, with its remarkably long and complete record of fire history, with its array of ur-landscapes abutting ultramodern ones, with its metastable bonding of human institutions like Bandelier and LANL, with its sudden eruption of landscape-scouring fires, as though the slumbering volcano had mated with plutonium and spawned Loki-like freaks to sport with the forest, the Jemez Mountains easily qualify as one of the geodetic markers of southwestern fire.

It's less obviously a cipher for the future. Each observing group sees its own agenda in the flames and its prophesied future in the postfire ruins. Are the eruptive fires agents of irreversible change? Or are they simply synthesizing the changes around them? Are the Jemez forests a model for the Southwest's sky islands, the beta version for the regional future of fire? Or are the mountains once again an exception, a place of spasmodic violence, its megafires the modern outflows of a megavolcano?

The temptation is strong, in particular, to make fire a subplot in the saga of global climate change. But it makes at least as much sense to see climatic warming as a subplot in the epic of fire history. When the Earth's keystone species for fire changed its combustion habits and reached into the Earth's past for fuel to stoke its craving for more firepower, it committed the planet to a new future of burning. Ancient and modern fires are still seeking a working equilibrium.

In a pragmatic sense it matters little what the ultimate cause might be. For the next few decades, perhaps for the next few centuries, the reality promises more of the same as the Jemez moves toward an end point over which we will have marginal control. We can't muster a counterforce to halt fires of this savagery—we can't bomb them away. But lots of smaller actions, added together, might dampen and deflect the damage. The energy released by the Las Conchas blowup might rival several nuclear bombs, but only over the course of days. So, likewise, a counterforce might soften the fires' eruptive power by altering the landscape over many years until it reaches a calmer angle of repose. That strategy might work, or it might not. But the no-action alternative is clear. It's to stand on the caldera, as at Cochití Canyon, and peer into an equilibrium of emptiness.

THE MOGOLLONS

After the West Was Won

ITS ISOLATION, PARADOXICALLY, is what brought the Gila National Forest to center stage in the drama of America's public domain. In 1924 Aldo Leopold argued to convert 750,000 acres at its mountain pith into a primitive area, the first in the nation. Forty years later the Gila Primitive Area evolved into legal wilderness and led to a program of fire restoration. This big paradox has led to a deeper one. If wilderness lay at the heart of the Gila's narrative, it was out of wilderness—by definition an acultural place—that the Gila established a durable culture of fire.[1]

As fire suppression segued haltingly into fire restoration, the Gila's combination of isolation and mass made it an obvious candidate for a natural fire regime. Any place that could absorb a two-week pack trip—Leopold's definition of a wild landscape—should be able to handle a barrage of summer lightning fires. The Gila Wilderness (and later, the bordering Aldo Leopold Wilderness) became one apex of America's triangle of natural fire projects. The High Sierra held fire within basins bordered by granite baffles and barriers. The Selway contained flames within its furrowed summits and ravines, its flaming whitecaps held by the seawall of the Bitterroot Range. The Gila offered a more inclusive isolation.

Its core was a massif of volcanic ranges that rose from the surrounding plains like a high island in the Pacific. Much of the external terrain was public land, or given to dispersed ranching. In 1970 Silver City had a population of 7,751. The nearest metropolis was Albuquerque, 125 air miles from Mount Whitewater. Across the Rio Grande the White Sands missile range had empty space enough to test missiles and accommodate the Trinity test site. A place that could accept the first atomic bomb tests could cope with nature's fireworks.

Like the Selway-Bitterroot or the Sierra's Illilouette Valley, fire on the Gila acquired its legend. But the task proved harder than expected: the transformation occurred because a committed group of fire officers made it happen. Like so much of the fire revolution that boiled out of the 1970s, success did not come with a handful of smoking snags or simple proof-of-concept free-burns. It came through patience and persistence, by institutionalizing the program, by handing it down through generations, by creating that rarest and most essential of prescriptions, a fire culture. It came by replacing the first glamorous flush of flames with the inglorious task of maintenance burning. It came not through the adrenaline rush of pathbreaking but from rebuilding and patching an aging trail. A golden age had to survive into silver, and then into bronze. After the frontier passed, life went on.

The great sagas of the fire revolution sing of a heroic era full of bold visionaries and daring feats in letting fire flare with the symbolic flash of a supernova. The great story of the Gila is not that it experimented with natural fire—that was all but inevitable. The story is that a natural fire program survived and quietly thrived.

The Gila is a classic patch of the Southwest, which is to say, of abrupt contrasts. Mountain and valley. Forest and desert. Drought and deluge. Mesa pueblos and metropolitan city-states. The relics of ancient peoples and modern sprawl: cliff dwellings from the 12th century look down upon a rootless automobile culture. Southern New Mexico adds an equally sharp contrast between two economies notorious for ruinous land use (mining and Texas ranching) with public stewardship and outright wilderness.

The seasonal rhythms of spring aridity and summer monsoon have an echo in the long cadences of human settlement and abandonment. The land inclines toward preservation even as it obliterates. The Mogollons are a sky island but with a New Mexico accent; not a fault-block basin and range massif, but the fossil relic of Miocene volcanoes that blew their summit away to leave a caldera, like a broken geologic skull with a few teeth remaining.

It's an ideal formula for fire. The Gila averages 358 lightning fires a year—the greatest of any forest in the national system. With or without people the place burns. But its actual regimens vary by how humans interact not only with ignition, by starting their own fires and putting out nature's, but with the tailings of combustibles that spill down the slopes and nestle into nooks and crannies like pinyon roots because people can both strip away the fine fuels and let the land overgrow. Over the summer, and across the millennia, the Gila's fires have come like the mixed rhythms of its rains: spotty, drizzling, and gully washing.

The gross pattern aligns with that of the Southwest generally. Fire is routine. It ebbs and flows as migrations and climatic waves amplify or dampen ignition. A study of the ponderosa pine region of the Gila Wilderness from 1633 to 1978 found that fires returned between 1 and 26 years with an average mean interval of 4 to 8. Because not all fires leave scars, these are minimum values. The geography of burned trees suggests that, in good fire years, the fires ranged widely and simmered and flared throughout the season. In the 1820s and 1830s the burned area dropped, probably due to the onset of a regional wet phase. Yet year by year the pineries of the Gila Wilderness simply burned.[2]

Then the cadences of fire and climate stopped in their tracks. With the suppression of the Apaches, the land opened to mining, railroads, logging, and ranching. What went around the region came to the Gila: the grassy fuels that had carried fire and sustained the regime were stripped away. What happened to its soils happened to its fires. The fires retreated like Mexican grey wolves and grizzlies, and were then hunted down after the lands were reserved as national forests. But the process of fire exclusion began 20 years before the U.S. Forest Service committed to full-scale fire protection in 1908. What the agency did was less to exterminate nature's fires than to prevent their restoration as the natural order suggested. From 1890 to 1978 burned area plummeted under the onslaught of full-scale American settlement.

That great oracle of the Southwest scene, Aldo Leopold, observed of southern Arizona in 1924—the same year he pushed through primitive status for much of the Gila—that "one is forced to the conclusion that there have been no widespread fires during the past 40 years." Pinyon and juniper woodlands had replaced grasslands; soil had washed away. "When the cattle came, the fires went and the erosion began." Yet Leopold as assistant regional forester actively approved and enforced fire protection. In a 1913 circular he "asserted" that "fire prevention is the most direct of all our activities, and hence also susceptible of developing the greatest relative efficiency." In 1920 he denounced as anathema such heresies as "light burning," which, he insisted, "could undo all the good the Forest Service has achieved in fire protection." Soon afterward he had to qualify that judgment. Speaking of a lovely trout stream, he concluded that "a century of fires without grazing did not spoil the Sapello, but a decade of grazing without fires ruined it."[3]

If the contours of its fire history conformed to the regional template, so did the Gila's history of fire protection. The Forest Service essentially doubled its holdings in 1907, reorganized in 1908, and under regional forester Arthur C. Ringland, pursued systematic fire control. Soon, trails, lookout towers, heliographs, telephone lines, tool caches, fire guards, and cooperative agreements were in place, and rangers appealed to mechanization to hold the line (by 1917 fire guards were driving motorcycles with sidecars stocked with fire tools). Even so, most of the countryside was too inaccessible and undeveloped. The qualities that made the Gila attractive as wilderness rendered it difficult to exclude fire. In the 1920s Henry Mullin could recall walking "15 or more miles into the fire camp area" for one blaze. What mostly checked fires was a landscape still recovering from its shattering contact with American civilization. As its firescape improved, the Civilian Conservation Corps (CCC) entered in force to impose an infrastructure and populate firelines. When they departed, the Southwest Forest Fire Fighter program, staffed by local pueblos and villages, filled their ranks. After the Second World War air tankers replaced biplanes; helicopters, pack strings; and a smokejumper base at Deming, per diem fire guards.[4]

Only very occasionally could conflagrations break out, as the landscape slowly recovered from its 19th-century scalping, as hillsides that had once been slicked clean and valleys reduced to boulders regained enough combustibles to carry fire. In June 1951 the McKnight fire, of mysterious origin, blew up, like some monstrous resurrection of a prehistoric beast, and burned 51,000 acres. The growing might of fire suppression, however, mostly kept pace with the returning pyric potential. More of the mountains transferred into legal wilderness—by 1980 roughly a quarter of the Gila's 3.32 million acres. The landscape of the index-defining McKnight fire now lay within the Aldo Leopold Wilderness.

On the eve of the Forest Service's policy reforms, the Gila boasted a detection system of 13 lookout towers, supplemented by aerial patrol. It had a "semicentralized" dispatching system for ground, helitack, and smokejumping crews. It had access to two air tanker bases. Its fire organization, modeled on California's, was fast, efficient, and comprehensive. Silver City resembled a miniature Missoula planted amid the foothills of the Mogollons. Yet its fire officers appreciated that the Gila had "one of the most complex wildfire problems in the Southwest" because "resource management practices" since the days of Leopold had "created a buildup of man-made as well as natural fuels."[5]

What mattered most was the land's rebound from the oppressive grazing that had crushed fire's ancien régime. Until the mid-1990s livestock continued even in wilderness. Like a boulder tumbled over a cliff, the shock wave of grazing had set off a debris slide still seeking its angle of repose. Some of that early shock had been felt immediately as surface fires were pounded into dust. Others, like aspen regeneration and the insidious spread of woodlands, rode a longer wave. The upshot was, fuels were returning to places that had lost them and were thickening on woodland and high forest sites that had never shed them. As that heavy hoofprint lifted, the land's health recovered, but that also liberated the small biota, which could power fire to work its ecological alchemy.

Fire officers needed some counter mechanism to allow fire to return without immolating the land. On the 50th anniversary of the Gila Wilderness they bet that an avatar of the once-maligned practice of light-burning might help calm the brewing tempest.

The Gila had natural advantages for a program of fire restoration—isolation, abundant dry lightning, a biota whose middle swath was dominated by ponderosa pine, a strong infrastructure for fire's management. So did many places throughout the West, and especially the sky-islanded Southwest, yet only the Gila succeeded in a big way. Most observers credit its fire staff as the reason.

They did it right. They started cautiously, they advanced incrementally, they worked within the existing programs to gradually nudge them into a wider arena and bolder experiments. They did not begin with manifestos, create parallel organizations for fire fighting and fire lighting, replace old-model fire officers with wildlife biologists and ecologists, or instantly relocate vast tracts of the wilderness into natural fire zones. They had credibility as firefighters: they never surrendered that capability. They moved into the unknown from the known. But they did move, and generation after generation of fire officers added to the cache of techniques, experience, and restored firescapes. The Gila was not the first national forest to experiment with natural fire—that honor belongs to the Bitterroot in 1972. Nor was it the first southwestern site; in 1970 Saguaro National Monument boldly proclaimed a program and even contributed a name for the practice, "natural prescribed fire" (others soon transposed that phrase into "prescribed natural fire"). But the Gila adopted it in 1975, institutionalized the practice, came to dominate U.S. Forest Service statistics, and was pushing the boundaries outward while Saguaro's program was imploding. Still, while the new policies allowed for fire's reintroduction, they did not compel it. The policy lit a torch. For the practice to spread, someone had to pick it up and touch the land with fire.

As the story gets told, the breakthrough came in 1975 when, on a field reconnaissance, Fire Management Officer Don Webb and a 22-year-old staffer, Lawrence Garcia, mulled over the proliferation of dog-hair thickets of ponderosa they saw spreading like cheat grass. How are we going to cope? Garcia asked. We're going to use fire, Webb replied, and you're going to write the plan to do it. That kind of resolve was not uncommon at the time. What made the Gila special was that it translated intention into practice.[6]

They began by expanding prescribed fire from slash piles to broadcast burns and by using new leniencies in the modified 10 a.m. policy to grant fires in tricky sites and out of season more room to burn. They allowed prescribed natural fires after the monsoon rains had come. When the Forest Service formally replaced the 10 a.m. policy in 1978, Webb and Garcia embraced its greater flexibility. They followed Horace's old adage, *festina lente*: they hastened slowly. They burned in the relatively benign zones first while they protected structures and the stray community. They burned in the fall, not the spring when the big fires raged, and so gained experience and confidence. They enlarged the range of prescribed burning. They added more terrain to prescribed natural fires.[7]

The process quickened after 1984 when Steve Servis became the fire staff officer. Like many fire officers of his age he had grown up with fire control, much of it on the Gila, where his father had been fire officer. He went to his first fire at five. He was using backpack pumps while still in middle school. He dragged a driptorch for a prescribed burn at 13. Born to the agency he was tough, capable, and committed to fire, yet, precisely because he had known southwestern fire all his life, he understood the need to reintroduce it in more controlled forms. He argued that the Gila had to align its program more fully with the new policies, and as a second-generation Forest Service officer, he could speak with the authority of shoulder-to-the-shovel know-how. The new policy allowed for variable suppression; for any given wildfire you could choose to control, contain, or confine. Servis and the Gila did all three, and eventually they learned to do them all at the same time.[8]

In the ponderosa belt the program contained, in the pinyon-juniper it confined, and amid the mixed-conifer it suppressed. Fire crews loose-herded rather than let-burned. They would burn out a ridge, and let the flames drape over the slopes to block a too-vigorous run. They would cut line and burn out to corral a flank. They would order retardant runs, flight after flight, to quell the occasional bad event. They were constantly on a flaming front. The acres added up. They passed relatively unscathed through the post-Yellowstone gauntlet that left so many programs as roadkill. A 1992 environmental assessment allowed the fire program to "go large." A year after the 1995 common federal fire policy, the Bonner fire became the first landscape fire in wilderness, the inflection event from prescribed natural fire to wildland fire use. Once again, while the Gila didn't set national policy, it was an early adopter, testing its limits, and inspiring (and shaming) the laggards. They were heady years.

The Gila cultivated that most precious of fire management's needs: they created a culture. Each subsequent fire officer came with solid fire credentials, and each stirred more ingredients into the stew. Gary Benavidez returned after 10 years running the Missoula smokejumper program.

On his watch some natural fires in wilderness were allowed to leak over into generic wildlands. In 2003 the "wildland fire use" (WFU) Boiler fire burned 58,000 acres, churning out of the Aldo Leopold Wilderness to race over Boiler Peak and establishing a new benchmark. That managed fire had exceeded the acreage of the suppression era's standard, the McKnight burn. Then came Paul Boucher; then Toby Richards and Gabe Holguin. They were attached to the place, not just an idea; when most retired, they remained in Silver City. It was possible to assemble 30 years of Gila fire officers around a kitchen table.

Its legacy became the anchor point of the program. That inheritance was visible in the biota: the previously burned landscapes began to shape the spread of new ignitions. But it was equally true institutionally: each new prescribed fire or wildland fire use was not a one-off, bet-the-farm risk. It built into and around its predecessors, part of a consistent, mixed program that fought, lit, and variously managed fires. Gradually, the burning moved into new terrain and across the gamut of seasons and effects. The Gila burned all the time. The Black Range district in particular became a prescribed-fire training academy for the region. It seemed that the "dream" of the founders was happening, that crews could light a fire in one place, fight fires in another, and manage a natural fire elsewhere all in the same day. The Gila's fire folk knew fire in ways that had not been true for a century.

Even the program's staunchest advocates, however, appreciated its limits. They would have to burn 100,000 acres a year to meet a 10-year cycle for ponderosa, and were achieving much less than that. They knew that maintenance burning was harder to sustain and that had to continue in perpetuity. They openly admitted the economic "challenges," that there was always money for suppression and only intermittently for restoration, even as studies documented the costs of "doing nothing." And they recognized the "political climate" that accepted escapes during suppression but not during prescribed burning. Their tenacity stemmed less from idealism than a gimlet-eyed understanding that suppression-only didn't work.[9]

Still, the impact was national. The Gila did for Forest Service natural fire statistics what Everglades and Big Cypress did for National Park Service prescribed fire figures: between 40 to 80 percent of total national acreage occurred within the Gila. When the Boiler fire moved outside legal wilderness into wildlands, the founding vision that the goodness of the Wild could, through fire, propagate beyond its inscribed borders seemed to be happening. Each new burn overlapped and intercalated with its predecessors, gradually sieving combustibles and sculpting a fire regime. The Gila went beyond proof of concept to become an operational exemplar.[10]

Fires returned to the Gila but with a character different from those that were rebranding much of the West. The 2011 season distilled that fact amid two contrasting New Mexican landscapes. In the Jemez Mountains to the north the Las Conchas fire reburned part of the scar left by the feral Cerro Grande prescribed fire and thoroughly nuked swathes of countryside into ash and dust. In the Mogollon Mountains the Miller fire ranged across 88,835 acres, scorching some patches for the fourth time, and left a landscape more resilient than ever. Longtime observers of the Southwest fire scene like Tom Swetnam rejoiced at watching the flames free-roam the landscape "like a wild beast" happily released "from its cage."

Yet a legacy can be a burden as well. You cannot replicate or restore an era as you can wolves. In 1893 Wisconsin historian Frederick Jackson Turner noted that the 1890 census had given up the effort to trace an advancing line of settlement, and then (with wild prematurity) he announced that the frontier had passed.

That declaration would have surprised the millions of frontier folk who homesteaded throughout the 1890s and into the next two decades; in fact, the heaviest volume of homestead patents occurred during the 1920s. The task of settling the land hadn't ended. Most of the hard work of filling in the empty spaces after the forty-niners, cattle drovers, and speculators had vanished had yet to be done. But that next generation could not appeal to the dazzle and mythic bravura of a frontier. They settled the West after it had nominally been won. Initial attack was over: what remained was the tedious but necessary task of burning out, patrolling, and mopping up. They struggled without the cultural aura attached to pioneering. It seemed all labor and no glamour. Owen Wister's *The Virginian* yielded to Hamlin Garland's *Sons of the Middle Border*.

That is not a bad model for what has happened on the Gila. The heroic age has passed. The easy mines have been worked out, the most fulsome pastures eaten down, the accessible timber hauled out, and the firescapes most amenable to restoration put into a new regime. Now it is maintenance burning, and tougher treks into woods for which fire's reintroduction will be much trickier, and all this in the context of punishing drought and that other legacy—of amassing combustibles, like clutter overspilling an attic—that made selective burning difficult. The easy-worked biotic placers of ponderosa pine gave way to the hard-rock mining of mixed conifer—a corporate, high-capital enterprise no longer accessible to the entrepreneurial prospector. The program moved from high-visibility symbolic fires to routine burning, and from a handful of fires to a functioning fire regime.

By now, too, the old infrastructure that had diffused fire knowledge throughout the Forest Service and had put smokejumpers, helitackers, and air tankers on rapid call-up was gone. Fire officers had to do more acres with fewer resources, with much less room for maneuvering, and with fewer buffers. There were more restraints from top management and more pushback from nonfire colleagues as fire costs threatened to consume the agency, like the Santa Rita mine swallowing the countryside into its expanding pit. As the Gila's fire management officer Paul Boucher put it in 2004, fire management was becoming all risk and no reward.[11]

Today, Silver City and Grant County still celebrate Billy the Kid, the Apache wars, and its heyday of mining. Yet the old sagas make little sense in an Old Town cluttered with art boutiques and the wintering homeless, and a countryside sprawling with retirees. This time the frontier really has passed.

So has the fire revolution and its early euphoria, along with many of the features that had made the Gila inviting as an arena for restoration.

The litany of what the euphemistic call "challenges" would stagger most forests. Brutal drought, lessened institutional capacity, tougher terrains and more maladapted woods, fiscal constraints that led to a national "emergency" reinstatement of the 10 a.m. policy for the 2012 season, even the success of a program that had treated probably 75 percent of the high-return ponderosa forests and left the marginal mixed conifer—amid such considerations it would be only too easy to let the maintenance go another year, or two, and leave the problem for someone else. When the budget shriveled and drought deepened, prescription burning plummeted from 18,000 to 20,000 acres a year to 5,000 to 6,000. When the program began, red flag days were an exception. Now fire officers assume the winds will blow hard, and look for breaks. Even as the fires run with those blasts, the fire program must tack against strengthening political headwinds.

Like the southwestern landscape, fire management on the Gila slapped old and new together and kept parts of everything. Here, history has mattered. Its preserved past granted the Gila a culture. If the wreckage of past times persisted, it was equally true that the triumphs were remembered as well. Today's fire regimes have built on both. That transfer of lore made the Gila a national icon. But generations that didn't leave could also become a nag and a burden.

And so it proved in 2012 when lightning kindled two fires deep in the Mogollons. Both were attacked, but so severe were the conditions, so volatile the fuels, so rugged the terrain, that crews were pulled out a day later. The fires merged into the Whitewater/Baldy complex and raged into the higher elevation mixed-conifer forests. They blew the top off the biota, leaving ecological calderas. A Forest Service retirees group investigated and issued its own after-action report. In the old days, they insisted, such an explosion would have been impossible: the ignitions would have been hit instantly and beaten back. The world they had worked to save from flames had been savaged, and they took it personally. NIMBY—Not In My Back Yard—acquired a corollary: NIML—Not In My Lifetime.[12]

In his 1948 novel *Fire,* George Stewart included a subplot about his ranger protagonist, Bart Bartley, who has a special attachment to a patch of old forest in a place called the Basin that he wants to protect. It clouds his judgment when the Spitcat fire spills over the district. In the end, his effort to spare the Basin causes him to lose both the fire and the woods. So perhaps it is with the effort to shield the dense forest of the mountains. They will burn. Their immolation is what regenerates the aspen whose colors make the peaks so photogenic every autumn. But those fires might come once a century, and a man could live a lifetime and never have to witness their incineration and slow renewal through the tangle of black locust, windfall, and sentinel snags. Critics could see the scene in terms of favored patches, not landscapes, and of fires, not fire regimes. They saw, and cherished, the shaded green of the mixed conifer they met when they first came to the mountains, not the new-growth aspen that would replace it. Yet fire management is a practice in which winning firefights can mean losing fire regimes.[13]

Outside those eruptions the evidence suggests the Gila is weathering the crisis better than most of the Southwest. Decades of steady burning of all sorts have granted a resilience denied other landscapes. The mountains are dappled with burns. The immense wilderness at the Gila's core means fire officers have a big box to contain fires that blow out bad patches. So, too, a program that has built swath by swath, year by year, has an institutional resilience few others can claim. Its fire corps has long sought to fight and light at the same time. "Fire is fire is fire," Gabe Holguin shrugs, "and we'll manage it."

That kind of continuity, however, can present a very different challenge than blazing trails into new country, and one that rarely attracts praise. Kudos go to novelty, not maturity. This may be a time on the Gila not for new growth but for survival. While its signature species, the ponderosa, will not regenerate amid deep drought and a fury of flames, established trees with deep taproots can survive. They can ride out the long waves of hostile climate and unfavorable politics provided they don't face too much stress from too many competitors. Historically, fire culled out that throng, adjusting population to circumstances. Too long without fire, however, and flame's reintroduction might only add to the strain because the resulting fires will not be those the landscape had evolved to accommodate.

What old and new management regimes might share is the value of persistence. In the formative years persistence meant steadily pushing into new land. Today it means holding on. Lose fire from a landscape, and it can be risky and costly to reinstate it. Lose a fire culture, and its reinvention will never reconcile completely with the lands it lives on. But those truths build on another. The Gila will have fire. That is a metric as constant as the speed of light. The issue is what kind of fires come and the character of those who must face the flames.

THE HUACHUCAS

Fire's Borderlands

THE HUACHUCA MOUNTAINS are not the tallest or the bulkiest or the most storied of Arizona's sky islands. No telescope adorns their summit. No historic tragedy or environmental controversy darkens their slopes. The Santa Ritas to the northwest boast the first experimental range. The Dragoons to the northeast contain Cochise's stronghold. Although Fort Huachuca was established among a chain of posts to contain the western Apache, it never acquired the notoriety of Camp Grant, the scene of a massacre, or of Fort Bowie, glamorized by Hollywood for its proximity to Apache Pass. The Huachucas are a middling mountain with a middling history.

Stand atop Coronado Peak, above Montezuma Pass, however, and you see what makes the Huachucas special. Everywhere is a landscape of borders. Between the Coronado National Forest and the Coronado National Memorial, the U.S. Forest Service and the National Park Service. Between high desert grassland, oak woodland slopes, and coniferous crests. Between wildland and private land, wildland and city, wildland and wilderness. Rise higher, and you could add Fort Huachuca, the Nature Conservancy's Ramsey Canyon Preserve, the San Pedro National Conservation Area, tiny hamlets becoming sprawling retirement enclaves, and ranches sloughing off their herding heritage like rattlesnakes their old skins and writhing forth as subdivisions. And within unaided eyesight to the south, there is the rusty ribbon of a fence that divides the

United States and Mexico. It's a landscape of borders, hard and soft, and of borders transgressed. Fire lookouts yield to Customs and Border Patrol (CBP) surveillance antennas, and the smokechasers of an older era to CBP agents pursuing smugglers.

The bedrock of Coronado Peak is breccia, a volcanic conglomerate of magma and shattered country rock. Its biotic cover oddly resembles it. The landscape is full of burnt oak trunks and resprouting brush, patches of severely and lightly scorched forests, a violent shattering and melding through the eruptive blast of megafire. The political landscape, too, is one of institutional fragments, not linked so much as slammed together. How to remake a fire breccia into a working fire mosaic is the purpose of FireScape.

What set the FireScape project in motion were two shocks, one slow and one sudden.

The slow shock took the form of serial court injunctions based on the Endangered Species Act (ESA). With the National Environmental Policy Act (NEPA) and ESA, asymmetrical legal warfare was possible with the Forest Service; and over most of the Southwest's national forests, the upshot was to shut down logging. That did not affect the sky islands, which lacked the volume and access to sustain more than a quick flush of cutting. But rare species abound in the sky islands—endemism is a defining trait—and threats can stall efforts at the thinning, masticating, and deliberate burning that are believed essential to fire management. The controversy reached absurd proportions as partisans debated whether there might (or might not) be ridgenosed rattlesnakes that might (or might not) be harmed by prescribed fires. There were other species in the shadows of stock tanks and riparian corridors—the Huachucas have 17 threatened and endangered species; and more unsettling, new ones like jaguars and ocelots were slipping north across the border. The quarrel between chainsaws and ESA spilled over and tainted efforts to restore fire.

The sudden shock was the abrupt appearance of monster fires that began to plague the Southwest. Amid this sprawling epidemic the sky islands were not immune; if anything, their insularity made them more vulnerable because a giant burn could sweep over an entire range. The

Rattlesnake and Horseshoe fires in the Chiricahuas showed the potential scale and damages. The new mutant fires could unhinge everything that made the sky islands ecologically valued.

But the sky islands had assets, too. There was no entrenched (or revanchist) timber industry to negotiate around. The mountains were each a confined entity: they were model case studies in island biogeography. Landscape ecology had evolved sufficiently to furnish some intellectual underpinning and a biological counterweight to an emphasis on singular species. Ancient geography and emerging science both reinforced the sense that small projects were a formula for failure, that only programs that spanned whole landscapes could succeed. In southern Arizona the natural unit for management was the mountain, the whole mountain, itself. The sky islands, too, had partisans more interested in their collective integrity than in their pieces.[1]

Among them was fire. Fire was fractal: it could burn single snags, patches the size of aspen groves or north-slope forests, and whole massifs. No endangered process act forced people to heed its habitat; but it needed only spark, wind, and the fuels that metastasized in its absence, and no court order or act of Congress could halt it. Ecologically, it deserved its place in the landscape as much as Chiricahua leopard frogs and Mexican spotted owls. It was far more fundamental to overall ecological resilience than northern Mexican garter snakes and the Sonoran tiger salamander. Its attempted extinction had unhinged the biotas of the sky islands far more than the extinction of any individual species, no matter how passionate their partisans. Only fire operated over the entire landscape, from sotol-studded grasslands to summits bristling with aspen, pine, and Douglas fir. Only fire could, in one vast exertion, change the conditions for all.

The critical insight behind FireScape, however, was not just the need for programs that embraced landscapes, which is to say, entire sky islands, but the need to refashion the legal landscape so that programs could operate across a patchy political scene, so that treatments of any kind could range intellectually and politically as widely as fire did. The vital need was to shape NEPA to accept landscape-scale programs, not just site-specific projects. It was not enough to reconcile fuels and habitats: they had to mesh with political jurisdictions as well. With that leap it might be possible to transform a fireshed, defined by the physics of fuel and fire behavior, into a firescape, characterized also by people and their institutions.

After the 2002 fires—for Arizonans, especially, the Rodeo-Chediski fire—a cold dread crept into a corpus of researchers who had worked on fire planning for the Coronado National Forest and the National Park Service in southern Arizona and who understood that the combination of more severe fires and a hobbled Forest Service could obliterate preservationist ambitions. The time for endless discussion and open-public commentaries had passed because a new generation of fires threatened to burn the amassing records of opinion as though they were so much windfall.

A core cadre of 35 from the University of Arizona, including students, along with representatives from the Nature Conservancy (TNC), met to decide on a course of action. They had some funding—not much—made available through the National Park Service's Cooperative Park Studies Program and from the Department of Defense (DOD) thanks to Fort Huachuca. And they did intend to act: they poured their labor into operational plans. By 2005 a basic framework was in place to reconcile species and fire. Each sky island had its own personality, its particular quirks, and so its own plan. At the Huachucas Brooke Gebow and TNC orchestrated the project. The BLM, the Audubon Society, and private landowners participated. Then the Forest Service stepped in to shepherd the Huachuca Area Fire Partners Fire Management Plan through the NEPA process. In 2009 FireScape received a finding of no significant impact. It was free to proceed.

By now conditions had worsened. The long drought had deepened. The Great Recession, and its political aftershocks, dried up federal funding. Prescribed fire stumbled over endless complexities until it could no longer be considered a default choice. Fires burned hotter, faster, less predictably. By the time approval had come many of the founding champions had departed. Still, there was some money for first-entry fuel treatments, a bequest to the Nature Conservancy brought half a million dollars that could be used for fire projects in Ramsey Canyon, DOD found some funds for fuel and fire on Fort Huachuca, and plans leveraged their resources by targeting strategic sites. Reinterpretations of federal fire policy in 2008 nudged fire officers toward a box-and-burn operation.

Most of what plagued the Huachuca Mountains was shared with the other sky islands. But its two hard borders, north and south, overseen by other federal agencies, changed the dynamic of Huachucan fire. To the south lie Mexico and U.S. Customs and Border Protection; to the north, Fort Huachuca and the Department of Defense. The USFS does fire work for both, and receives contributions from each that it would not otherwise have. The southern tip of the massif contains the Coronado National Memorial, overseen by the NPS. For the Sierra Vista ranger district, most of the remaining mountain lies within the Miller Peak Wilderness. Much of what is left is WUI either embedded in side canyons like Ramsey and Ash or abuts the mountain's grassy flanks like Sierra Vista and its exurbs. That doesn't leave much land for maneuvering.

But it emphasizes the power of Fort Huachuca to alter the larger setting. Prior to its establishment the mountain showed the fire regimes typical of the other sky islands. Most of the landscape burning occurred with early season fires prior to the monsoon. It's not known—may not be knowable—how many were set by people, but more spring fires have anthropogenic rather than natural causes, and certainly recent decades point to human fire. The chronicle of stand-ages and scarred trunks suggests that fires burned every 4 to 10 years in the conifer belt, more rarely along the crest, and more frequently among the grasslands and oak woodlands. That rhythm broke abruptly in 1877 when the army arrived to stay. The military presence ended the migratory movement of peoples as an ignition source. The fort's very existence created markets for timber and beef, which chewed up the woods and grasslands, and sparked mining, all of which shattered the ancient rhythms of burning and the capacity of some early-season fires in some seasons to creep and sweep over most of the mountain. While monsoon storms continued to kindle fires, they spread poorly. Over the next century only one truly landscape-scale fire occurred, in 1899, likely feeding on logging slash and which, paradoxically, ended logging operations; an aftershock burned in 1914. Not until 1977 did large fires return.[2]

The fort endured, and by the latter 20th century its environmental impacts morphed from cutting and trampling to urban sprawl. Sierra Vista evolved from a sutler post to the major entrepôt of Cochise County.

Though military funds rose and fell with wars, they were more consistent than the funding given federal land agencies. As the army's primary facility for military intelligence training and testing, the fort thrived in an era that weaponized electronics and moved into cyber conflicts. As the original fort had encouraged settlement as a haven from Apaches, so it now provided a safe haven from a renegade economy. The town sprawled, and spawned others like Huachuca City and Hereford until the eastern flank of the Huachucas looked like a hundred other front ranges throughout the West. Few ranchers had welcomed fires, though they would tolerate premonsoon burns since imminent rains would turn them into verdant pastures and cut back brush encroachment. Exurbanites saw no saving graces from burns of any pedigree. A service economy based on modern sutlers, retirees, and recreationists wanted nothing to do with fire or smoke.

Still, the fort controls the northern half of the Huachucas. It is also subject to environmental legislation (through the 1960 Sykes Act); but the dynamic of management is different, and it upsets the familiar dialectic that has pitted the Forest Service against environmental activists, an Evil Empire against ESA insurgents. If the fort complicated Forest Service administration, it also created a potential partner. For FireScape the two big agencies (along with the smaller NPS) had to syncopate operations if fire was to be restored. Unlike the Four Forests Restoration Initiative, which still pinned its financing on an illusory wood-products industry, Huachuca FireScape had DOD to assist, and in a smaller way, Customs and Border Protection.

Then the Monument fire remade the facts on the ground.

Throughout the 20th century fires had been many, but big fires few. A hundred years after two companies of the Sixth Cavalry had pitched tents that became Camp Huachuca, a fire burned over Carr Peak. Another swept Pat Scott Peak in 1983. But the 2011 Monument fire burned most of the Huachucas south of the fort.

It began on June 12 near the Mexican border—exact cause unknown, but certainly human, and likely from border crossers. It bolted through the Coronado Memorial incorrectly identified by Forest Service dispatchers as a "monument," thus lending its name to the burn. It moved briskly through montane grass and woodland, scorching nearly all of the memorial save its riparian niche, before pushing north and upwards into the Miller Peak Wilderness. From there it blasted out through side canyons to the east. Ash Canyon. Stump Canyon. Hunter Canyon. Miller Canyon. Carr Canyon. Winds rushed over the mountain and drove down the gorges like chinooks, leaping across the four-lane Highway 92. The Hunter Canyon outbreak was the largest and accounted for most of the 79 structures (59 of them houses) destroyed. The flames even recrossed the border south into Mexico. Before the runs stopped, the Monument fire had terrified the east slope communities and burned 38,000 acres.

That was by area a third of the Huachucas, but the fire was dwarfed by the almost order-of-magnitude larger Horseshoe II fire and the 15-times larger Wallow fire, both burning in Arizona at the same time. The common conclusion was, whether or not plans incorporated the latest science on fire behavior and landscape ecology, whether or not those plans had approved environmental assessments, whether or not treatments have been applied, fire was returning to the mountains. If people didn't do it, nature would.

The Monument fire reset the landscape. Fort Huachuca had escaped, and began expanding fuel treatments. The mountain might serve as a proving ground for gadgets destined for Afghanistan, but testers and trainees didn't want incoming fire in the mix. DOD had options the Forest Service didn't, and could hire others (including the USFS) to slash and burn 5,000 to 6,000 acres a year. Meanwhile, the Sierra Vista District found itself overwhelmed with expanded mining claims at Patagonia, continued federal downsizing, and the border with Mexico. A visitor was unlikely to see a Forest Service ranger, but could not avoid swarms of border patrolmen. The Monument moment taught exurbanites about the risk they faced, yet paradoxically made them gun-shy of any fires, including prescribed. FireScape steadily dropped in priority. The fire it was designed to forestall had already occurred.

The borders loomed larger than ever, and they promised some weird and destabilizing synergies. Mexican species were moving north along with drugs and migrants. Jaguars were sighted in the Chiricahuas and the Santa Ritas. Ocelots were slipping into ranges held by wildcats. Northern Mexican garter snakes slithered under the fence. There was precedent for newcomers to naturalize—most of the waving grasses that clothed the lower flanks and valleys were Lehmann lovegrass, a South African exotic imported for pasture improvement before becoming an invasive. The critical fact for fire managers was that immigrant species were establishing themselves outside the scope of the NEPA-approved plans and threatened to force the agencies into a new round of consultations. Given the legalistic character of American society, species protection and fire restoration found themselves in competition.

The originating FireScape insight had been correct. But timing had favored big fires, not big programs for fire management. Each of the sky islands has its own informing eccentricity, but the Huachucas are notably about borders. About borders that need to blur so that fire can be managed on landscapes as fire understands them. About borders that need to be stiffened so that problem crossings might be abated, that a fire program need not rely solely on the funding and standing of any single agency, that it is possible to promote fire where it is needed and keep it out of places where it is unwanted. The Huachucas show that border blurring can introduce problems as well as solutions and that borders change with the climates of shifting monsoons, unstable economies, and ideas.

The belief that America might get ahead of its fire scene seems increasingly unlikely. Its fire borders are defensive, not lines for advance. The best guess is that managed wildland fires, not managed wildlands, will underwrite the future of Huachucan fire.

THE KAIBAB

Friendly Fire

If I were the Prince of Darkness, I could not have devised a better way to destroy the Kaibab Plateau.

—WALLY COVINGTON

WALLY COVINGTON, PROFESSOR, restoration ecologist, and a man who has been around burned woods all of his career, is walking through the still-raw scar of the Warm fire that in June 2006 blistered, seared, scorched, and not occasionally incinerated his much-loved Kaibab forest. The fire was kindled by lightning, allowed in the name of naturalness to simmer on the land, and then boiled over with a savage ferocity that turned 58,000 acres to smoke and white sticks, wiped out nine nesting reserves for the northern goshawk, shut down the only roads to the plateau, including one to the Grand Canyon's North Rim, threatened a substantial chunk of the remaining habitat of the flammulated owl and endemic Kaibab squirrel, may cause a quarter of the old-growth ponderosa pine to die, promoted gully-washing erosion, and rang up suppression costs of $7 million. To help pay those bills the Forest Service initially proposed to salvage log some 17,000 acres of the burn, which has sparked promises of monkey-wrenching by local environmental activists. When trotted out before cameras after the blowup, the district fire staff officer, Dave Mertz, declared that if he knew then what he knew now, he would have made exactly the same decisions. Fire belonged on the land. This was an inevitable fire, a necessary fire, a good fire. Wally Covington thought it testified to ideology gone mad, and had the temerity to say so and the clout to be heard.

I was there because I wanted to come home. Forty years before, in June 1967, I had begun my own career in fire on the North Rim. Only five years previously had the opening salvo in fire's great cultural revolution sounded. By my second summer the National Park Service had rewritten its policy to encourage more fire on its lands. I wanted to see what that revolution had wrought.

When they began, the revolutionaries were one, united against a sharply etched villain. They all detested the paramilitary swagger and waste of fire control, and scorned its justifications. It had become, they denounced, a law unto itself, divorced from the purposes of protected reserves. They thought that some fire—more than then existed—belonged on the land. They believed the real problem was fire control, and that controlled fire—what they termed prescribed burning—was smarter, cheaper, safer, and more ecologically benign.

It was an era poised for change, and the first boulders that rolled over the cliff in 1962 set off a landslide. The Tall Timbers fire ecology conferences, the inaugural prescribed burn by the Nature Conservancy, the Leopold Report for the National Park Service, the Wilderness Act—the war on fire soon confronted multiple insurrections. One came from a public aroused by environmental concerns and enthused by a wilderness ethos, one from among scientists newly sensitive to fire's ecological services, and one from nature, which made increasingly apparent that fire's removal could be as ecologically powerful as its introduction. It became difficult to argue that natural fire did not belong in natural areas, and if good there, that its goodness should not be distributed widely through the proxy of prescribed burns. Very quickly, intellectual opposition collapsed. The revolutionaries had needed to do little more than find a bullhorn through which to speak their minds.

William Wallace Covington, regents' professor at Northern Arizona University (NAU) and director of the Ecological Restoration Institute, was a child of the Sixties. Born in 1947, the middle of three sons, he was raised in Wynnewood, Oklahoma. His father, a jack-of-all-trades from prize-fighter to radio announcer to barnstorming pilot, was above all an ardent woodsman, a one-time forester, who got himself and his sons where they could camp, hunt, and fish as often as possible. Sundays were spent, alternately, at his mother's Methodist church or his father's church of the woods ("mostly my dad's," Wally recalls). For scripture, his father introduced him to Aldo Leopold's *Sand County Almanac.* Then, when he was seven, his father died, and the family later moved to Grand Prairie, Texas, until he enrolled at North Texas State. Shortly afterward, he married.

Several years after his father died, his mother, he recalls, came down with a life-threatening cancer. That nudged him into a premed undergraduate program, research in oncology, and a year at the University of Texas Medical School where he pursued a dual PhD-MD program. He quit when he found he could not take the emotional pounding that dealing with terminally ill patients demanded, those who were beyond restorative care. He took a year off and taught at Gallup, New Mexico, where he became an all-purpose activist for the causes of the day. "I was basically a hippie." His contract was not renewed. He returned to school for a master's in biology at the University of New Mexico, and then went to Yale's School of Forestry, where he did fundamental work on soils and logging. By the time he graduated with his doctorate, he had accepted a post at Northern Arizona University in Flagstaff, Arizona, and in 1975 made his first, epiphanal visit to the Kaibab Plateau.

These were landscapes dominated by ponderosa pine and fire. But they were also landscapes deeply ill from past overgrazing that had strip-mined the understory grasslands, from the logging off of old-growth that had broken the forest structure, and from the exclusion of free-burning fire that left those landscapes to gag on their accumulated debris. They were, in brief, landscapes not unlike the cancer patients who had attracted and exhausted him; but this time a cure was possible, or better, a prescription for prevention. The research he conducted under Forest Service contract assumed a reintroduction of fire as the means to restored health.

By the mid-1970s that was the collective wisdom of the day: fire had to reenter the landscape. The tricky issue was how. The National Park Service formally revised its policies in 1968. The Forest Service stepped through a sequence of half measures, allowing some wilderness fires in 1972, publicly converting in 1974 to the doctrine that fire management had to serve land management, and adopting a formal policy of fire by prescription in 1978. But ideas were easy, implementation tough. The American public became rudely aware of what was happening when Yellowstone burned through the summer of 1988. Apologists cleverly framed the ensuing outcry by debating whether fire belonged in Yellowstone. That was easy; of course it did. What they avoided was the gist of the operational issue, how and at what cost and under what social compact should fire belong? Even as the NPS burned up $130 million to $300 million (no one knows exactly how much) while failing to control the fires, the Park Service and its apologists managed to skip over the point where the philosophical rubber hit the road of real-world ecology. It was, it claimed, only doing what came naturally.

It was exactly this issue that the fire community has never resolved within itself. The revolution, like a bar magnet, had two poles that held the fractious particles within a common force field. The poles were bicoastal. One resided in Florida, focused on the Tall Timbers Research Station and the charismatic Ed Komarek; a sense of fire as used on private land, fire as historical and cultural, fire as a means to promote biotic assets, whether longleaf pine, bobwhite quail, or open-woods cattle. The other pole centered on the national parks of the Sierra Nevada, with its intellectual anchor in the University of California-Berkeley and its focus on public lands, and for its prophets such wildlife and range professors as Starker Leopold and Harold Biswell. The Florida pole wanted fire in the hands of people; the California, as far as possible left to nature. Behind the wilderness model was the expectation that, while prescribed fire might be necessary as an expedient, the agencies would ultimately surrender their colonial oversight to the indigenous processes of nature. Prescribed fire was an expedient, to be succeeded by natural fire as possible.

Of course fire control never collapsed, and as exurbs have spread like a contagion, fire suppression has regerminated like scrub oak, its roots all the stronger for the hacking it received. But it lost its sense of singularity and inevitability; fire protection was no longer the sole point of fire agencies. And although the debate continues to be dressed up in a Smokey Bear costume for public display, with the choice presented as one between fire's exclusion or fire's inclusion, the real decisions are among the ways and means of reestablishing fire. Fire's great cultural revolution had begun with a fire in the minds of men, and for the federal agencies that discourse had ended 30 to 40 years ago. What has remained is to put those fiery ideals on the land.

Unfortunately, putting fire back into the land has proved more daunting than taking it out. Shortly after arriving in Flagstaff, Wally was working with Forest Service researchers keen to reinstate fire. They believed that reintroducing fire would be enough to clean out the clotted understory that choked the land like woody plaque. Plots were laid out, burned, and assessed. But everyone knew the results could only be good.

After a few years, however, the field trials showed outcomes that were the exactly opposite of what had been predicted: loosed fires had killed few of the young trees without burning them up, while slow-cooking fires had girdled the base of the old-growth ponderosa, two-thirds of which according to Wally's survey died over the next 10 years. This was not what agencies wanted to hear. The Forest Service had just completed its painful conversion away from fire's suppression to a doctrine of fire by prescription. Wally's agency cooperators demanded he surrender his data on old-growth tree mortality since his work was under contract and hence the property of the Forest Service. The results, while obvious to anyone who visited the sites, were not published until 25 years later.

The solution was to establish his own program, which he did in 1978. In 1992 he laid out plots on never-logged lands in the Fort Valley Experimental Forest outside Flagstaff. These would create a standard, a natural referent, the sort of desideratum that Wally had derived from his reading of Leopold. The next year Wally received a big National Science Foundation grant to create his baseline, which soon documented that the current forest bore little relation to its presettlement predecessor. That forest had resembled a savanna with small groves of yellow pine dappling glades of grasses and forbs and washed by frequent fires across the surface. An onslaught of sheep wrangled those flames out of the land. Overgrazing ended fires, which found nothing to burn; this allowed reseeding to inedible woods, which grew in stunted thickets, starving everything on the site and encouraging the occasional ferocious fires that could wipe out even the canopy of ponderosas. A pavement of pine needles buried the site's biodiversity. Outside the Fort Valley plots, where logging had felled old-growth ponderosa, the forest structure had further degenerated. The landscape was deeply ill, and without aggressive treatment, it would die. It would shrivel during prolonged droughts, become sick with beetles and mistletoe, and ultimately burn up in an escalating fever of windy holocausts.

Wally concluded that fire alone could not reverse this decline, for fire could only act on what existed: messed-up woods were likely only to encourage messed-up fires. Rather, the solution was to first restore the structure of the old forest by thinning out the intrusive conifers and then applying fire. Function would follow form, and preliminary experiments seemed to bear out his belief. At this point "restoration" acquired a historical dimension to Wally, for it was the forest as it existed before the shattering blows of settlement that seemed to furnish a reasonable target for restored health. Cutting the small stuff and getting surface fires back on the land would spare the old-growth yellow pine from extinction. While he shared with environmentalists a passion for Aldo Leopold, Wally read the Leopold of the land ethic and the Leopold who healed a Wisconsin farm deeply scarred by settlement, not the Leopold of the wild. Out of this experience evolved what became known as the "Flagstaff model."

As the plots matured, drought brought a wave of wildfires to the region, scouring out large swathes of ponderosa forest that had occasionally torched but not known sustained crown fires. One after another, like sections of a giant fault line rupturing in quirky sequence, patches of ponderosa forest around Flagstaff erupted under conditions quite different from their evolutionary heritage, powered by drought and the explosive mantle of young conifers that had grown up like an immense shag rug under the old-growth canopy. To Wally, the scene demanded quick and massive remedial action. A century of abuse had left the forest too enfeebled and vulnerable to recover from the kind of trauma such fires inflicted.

In 1994 Wally wrote a seminal paper on his conclusions to date. That summer off-the-charts wildfires shocked the national fire establishment, sparking a new common federal fire policy and alarming thoughtful observers about what the future might bring. Secretary of the Interior Bruce Babbitt, from an old Flagstaff family, became interested, enlisted Arizona Senators Jon Kyl and John McCain to provide political leverage, and had the Bureau of Land Management make available some land at Mount Trumbull in the Arizona Strip country on which to establish field trials. (It probably helped that Mount Trumbull was so remote that it might be considered the Area 51 of fire research.) The Mount Trumbull experiments scaled up the Fort Valley plots to something approaching operational acreage. In 1997 Wally's efforts acquired a surer institutional identity with the creation of the Ecological Restoration Program at NAU, later upgraded into an institute, with various funding from Interior and then a congressional set-aside through the Forest Service. The Flagstaff model began to get national attention.

Which meant it also attracted national critics. Fire's great cultural revolution had been a bubble in a larger pot aboil with enthusiasms, a kind of environmental Great Awakening, which for many had become a kind of secular religion. It had its magical icons, among them reintroducing the wolf and dismantling Glen Canyon dam; but perhaps nothing commanded more practical symbolism than to stop logging on the public lands. Particularly with the use of the Endangered Species Act, that was slowly happening. Now in the name of fire protection, ostensibly to spare old-growth ponderosa from incineration, Wally Covington, a retread forester, was proposing to bring a species of woods-product industry back on the land. Critics suspected that the Flagstaff model was not about reintroducing fire but about reinstating chainsaws. When the Flagstaff model was cited in the National Fire Plan authorized in 2000 and the Healthy Forests Restoration Act of 2003, Wally's program became visible, powerful, and suspect. During the Southwest's record 2002 fire season a color photo of treated Flagstaff forest appeared on the front page of the *New York Times*.

The polarizing of American politics meant compromise would not be easy. What Wally saw as science-based preventive medicine, others saw as forester-inspired quackery. What hard-core wilderness proponents saw as deference to nature's way, Wally saw as surrogate religious sentiments—ecology as theology. Had he known what was to follow he might have traded his father's Leopold for his mother's Methodism, and cited Paul's epistle to the Hebrews, "For our God is a consuming fire."

The fact was, fire's reintroduction simply didn't happen on the scale desired. Setting fires was complicated, and letting fires free-range proved less widely suitable than expected. Meanwhile a decade of celebrity conflagrations roared across TV screens, usefully timed with national election years and unhelpfully thrusting suppression back to the forefront as an exurban fringe met a burning backcountry. Firefighting costs went ballistic, agencies were traumatized by an outburst of fireline fatalities, and the argument that prescribed burning alone could set matters right seemed too complex and long range to meet needs. Besides, those who had witnessed the revolution wanted to see its results. They wanted regime change. Impatience grew.

The bicoastal split became more pronounced. Agencies began to set numerical targets for burning, although most of the acres occurred where they always had (the Southeast), not where they were most loudly boosted (the West). But the West had something the South didn't: vast tracts of uninhabited land, some in legal wilderness, some just public wildland. This suggested that naturally ignited fires could be left to roam under designated conditions. This had always been a goal of the founders, but the early experiments attempted a synthesis of the two traditions, what became known as the "prescribed natural fire." The practice began cautiously amid several spectacular failures in the late 1970s (though not advertised to the public), and reached its apotheosis in the 1988 Yellowstone outburst. After that blowout, the expression began to disappear until a new avatar appeared, "wildland fire use," a term as generically meaningless as "prescribed natural fire" was wonderfully oxymoronic.

A WFU fire was a burn that advanced agency objectives for management. In principle it seemed too good to be true. It promised to be cheaper, safer, and more ecologically wholesome. As with prescribed fire, it required preapproved plans and had rules, and as with suppression, a WFU required suppression crews in reserve. But as long as the fire stayed within the domain allotted for it, it was doing nature's work. It returned fire's agency to nature. And not least, it absolved institutions of liability. They had not started the fire, nature had. Nature made the decisions, nature determined what would incinerate and what would survive. No one could be sued. There were no permits required for smoke. There would be no haggling with environmental groups over a human presence. You would outsource fire production to nature as companies might outsource manufacturing to countries with less-onerous taxes or labor laws. A small fraction of such fires, left to loiter, might go looking for trouble and bolt away, burning up what years had conserved; but by then the fire was a wildfire, a bureaucratic metempsychosis that allowed its costs and consequences to be borne by suppression.[1]

Yet suppression still provided the organizational template. From the beginning, some reformers had tried to create a parallel organization, seeking the same kind of funding, comparable crews, an equivalent public acknowledgement; they lacked only a parallel Smokey Bear. Now, a similar logic of success appeared. In the past a failed initial attack had become an argument for further, fuller suppression, for a better initial attack could have caught the fire when it started. Now, a failed WFU became an argument for fire's more comprehensive reintroduction. More burning, a wider geography of WFUs, could have dampened fuels and prevented an escalation to conflagration. A rogue WFU was regrettable, but did not alter the fundamentals. The failure lay in execution, not conception. Besides, in any conflict there will always be victims of friendly fire.

To some this might look like the ideological equivalent of money laundering. But to many, raw with impatience over decades of dawdling with fire's reintroduction and alarmed as fire suppression consumed more and more agency budgets, an expanded program of WFUs seemed a slick solution. It could get the burn out the way the old logging-driven Forest Service had sought to get the cut out. In 1996 the Kaibab National Forest began to plan for prescribed natural fire (PNF). The claims advanced were that PNFs would be "low-intensity burning" of a sort "not currently available" on the forest, that the PNF would occur "under specific guidelines and favorable climatic conditions," and that "the use of prescribed natural fire will assist in changing the current situation of infrequent, high-intensity fires to frequent, low-intensity fires." Unfortunately the proposed guidelines, known as prescriptions, did not include an irony index. In 2000 the Kaibab approved the PNF program, now renamed Wildland Fire Use. The environmental assessment concluded that the program would have "No Significant Impact." The Warm fire began as a WFU.[2]

The Kaibab Plateau is an outlier; the southernmost of the High Plateaus, the northernmost of the southwestern sky islands, the easternmost of the staggered plateaus through which the Grand Canyon is excavated. It appears like an inverted saucer, its low slopes disguising its height (over 9,000 feet). In Paiute it means "mountain lying down." Viewed another way, it resembles an altar.

It is a place that makes ideas seem original, then obvious, then too complicated to understand. The canyon is one example. Upon discovery, it quickly became a testimony to fluvial erosion; but after a century of inquiry over just how the Colorado River had veered west to cross the grain of the plateaus and make the excavation possible the mechanics remain unclear. In the 1920s, after hundreds of predators had been killed by government hunters, the mule deer herd "irrupted," overwhelming the woods. The browse line that ringed the plateau became as celebrated in conservation circles as clear-cuts in the 1960s. Aldo Leopold evoked the outcome by suggesting it looked "as if someone had given God a new pruning shears, and forbidden Him all other exercise," and then made the Kaibab deer herd an exemplar of game mismanagement. Removing predators had disequilibrated Kaibab ecology; Leopold concluded that such irruptions posed a greater threat to the forest than anything save fire. Yet the reality of that fabled irruption, much less its mechanics, is now in doubt, and its meaning garbled.[3]

Now, perhaps, it is fire's turn. On the Kaibab consensus views have a way of becoming burnt offerings.

Wally Covington first visited the Kaibab in 1975, and like so many others, found himself enchanted. Here was a place that might be spared the worst of the coming conflagrations; the Kaibab would be the apex of a management triangle that stretched from Flagstaff to Mount Trumbull. Until the 1990s only one large fire had blasted over the plateau, the Saddle Mountain Burn of 1960, which had begun in the park before ripping across the boundary. Since then there had been 300 acres burned in a wildfire on Powell Plateau, and two 1,000-plus-acre burns on the North

Kaibab National Forest. While the absence of fire was of course the problem, big fires had not already gutted the woods. In 1997 Wally and his colleagues at NAU commenced studies with Grand Canyon National Park to reconstruct fire history, similar to what he had done at the Fort Valley plots. Those data would provide a base level to determine what preventative treatments might be needed. Then matters stalled.

Part of the problem was that Wally had, in the eyes of critics, overreached himself. Convinced by the merits of the Mount Trumbull trials, he campaigned to extend those techniques into bona fide wilderness nearby. That caused pushback. Wilderness should be left to the wild. Besides, intensive treatment was expensive; ultimately it could only work if some market existed for the debris—small wood lumber, biomass energy, pulp, something. That looked even more like a wood-product industry in camouflage.

By now, too, the park had decided on other means. Flush with money lavished on the National Park Service in the aftermath of Yellowstone's summer of fire, the Grand Canyon finally got serious about a fire program, adopted a new fire plan in 1992, and began aggressively putting fire back into the land. Prescribed burns and WFUs began to blanket the woods. One fire, Bridger-Knoll, left for observation, bolted free and burned into the national forest and across some 50,000 acres. Another prescribed burn (the ominously named Outlet fire) went feral, forced the evacuation of the North Rim, and even skimmed over and around the canyon rim, reburning Saddle Mountain. (Revealingly, it started the day before a Park Service burning crew lit the fire at Bandelier National Monument that scoured out Los Alamos.) Probably 95 percent of the area burned since the park's creation in 1919 has burned in the 15 years that followed the 1992 plan. (For the Kaibab as a whole, roughly 90 percent burned between 1994 and 2006.) Whether or not biotic goals were being advanced, the park was getting the burn out—getting a mix of large-area fires, some of considerable ferocity. It saw little need for Wally's prescriptions. It had little stomach to slash down thousands of small-diameter trees that could only infuriate environmentalists, an undertaking it deemed unnecessary and expensive. It was finally getting fire on the land.[4]

Wally was sympathetic but dubious. This looked like faith-based ecology—sprinkle fire like pixie dust and everything will turn lovely.

The park had an expensive, well-staffed program sharply attuned to fuel loads and potential fire behavior. But someone like Wally concerned with old-growth ponderosa knew that even surface fires could kill a significant fraction by slow-burn girdling around the accumulated debris at their base; the tree died a year or two later, well after any postfire surveys. That had not happened historically because frequent surface flames had flushed away the needle cast every few years; now a century of compacted biomass was available to slow cook roots. Yet the park had no monitoring program in place to measure such biotic consequences. Nature would take care of itself. These ignitions, Wally insisted, were not natural fires; they acted on lands profoundly disturbed for a century; they might well be indistinguishable in their ecological consequences from those that the agencies had warned would devastate the public domain and that had helped jar loose billions in federal funding to prevent. Still, he regarded the Kaibab overall as redeemable. Then came the Warm fire.

The Warm fire started from a lightning strike on June 8, 2006, a few miles south of Jacob Lake, between Highway 67 and Warm Springs (which gave the fire its name) and could easily have been extinguished with a canteen and a shovel. The North Kaibab instead declared it a wildland fire use fire, and was delighted. Previous efforts with WFUs had occurred during the summer storm season and had produced small, low-intensity burns that "did not produce desired effects." The district had committed to boosting its acres and had engineered personnel transfers to make that happen. Since large fires historically occurred in the run-up to the summer monsoon season, when conditions were maximally hot, dry, and windy, the Warm fire promised to run up the desired acres.[5]

For three days the fire behaved as hoped. On June 11 the North Kaibab district requested a fire use management team to help run the fire, which was now over a hundred acres and blowing smoke across the sole road to the North Rim, and so required ferrying vehicles under convoy. On June 13 the fire use team assumed command of the fire. Briefings included a warning that the fire could not be allowed to enter a region to the southeast that was a critical habitat for the Mexican spotted owl.

The "maximum management area" allotted for the fire was 4,000 acres. That day the fire spotted across the highway, outside the prescribed zone. The fire team and district ranger decided to seize the opportunity to allow the fire to grow and get some "bonus acres." The prescribed zone was increased; then increased again as the fire, now almost 7,000 acres, crossed Forest Road 225, another redline border, which would permit the east-trending fire to enter swathes of pinyon-juniper that the North Kaibab had long sought to burn off. The assumption was that the fire would self-extinguish as it combusted through that dwarf forest and, falling off the plateau, entered sparse shrublands. On June 22 the fire, now roughly 11,000 acres, did abate. No full assessment of conditions, as required by the approved wildland fire implementation plan, was conducted.

The fire rekindled, moving east as hoped and north, which threatened the highway to Jacob Lake. On the morning of June 24 it was 15,000 acres, and the district ranger decided to order midlevel fire suppression help (a Type II incident management team) in case the fire broke loose and compromised the road or the complex of campgrounds, inn, and ranger station at Jacob. Forecasts called for developing northeast winds. But the conditions that favored fire on the Kaibab also favored fire throughout the region; there was in particular the tricky Brins fire, with massive media attention, that threatened Sedona. There was precious little left to commit to a WFU fire that might or might not need help in the remote Kaibab. A fire simulation model projected only modest growth. The model, however, was fed data from June 12, which bore little relationship to current or anticipated conditions, and it predicted fire behavior typical of surface burns, not crown fires. Its forecast bore no correspondence to reality.

On June 25 the fire blew up. The sought-for crown fire through pinyon-juniper lofted a convection column that evolved into a fire-generated thunderhead, a pyrocumulus. When the plume collapsed, it sent winds rushing downward, exactly like a thunderstorm but without a particle of moisture, a dust storm of flame. Those winds drove the fire south, spreading like a toxic cloud spilling out from a volcano—a cloud of combusting gas that flowed through the woods and across nominal barriers. It seared mature stands of aspen, it poured over meadows, it fried forests of mixed conifer. Terrain and turbulence meant that some patches were spared or lightly scorched. But it is unlikely that the Kaibab

Plateau had ever witnessed something on this scale and savagery. When the rush ended on the evening of June 27 the fire stood at 58,640 acres.

Friendly fire.

This fire really hit me hard.

As Wally walks through the charred landscape a year later, he wonders if the Kaibab can still be saved. The 1996 Bridger-Knoll fire, left for observation below the rim, had burned 52,000 acres to the west; now the Warm fire has gutted the center and northeast; and the park has imposed a rapid series of burns, some intense, to the south. Within a decade the fire regime of the Kaibab Plateau had inverted. The WFU program had promised to replace unnatural, damaging, high-intensity fires with natural, benign, low-intensity burns. In fact, it had replaced decades of small, low-intensity burns, held by aggressive suppression, with an eruption.

Wally wonders "how much is left to work with," if there is enough to save—if the scale and suddenness of the shock leaves sufficient flex in the land to warrant further preventative measures, or if the Kaibab will be left to sort out its own future. He points to a cluster of old-growth pine. "They'll be dead in a year or two." This was a landscape in rehab, not restoration. The Warm fire, he says, "really took the wind out of my sails."

More than a change in wind was a change in climate, the one that mattered most, the climate of opinion. Wally Covington found himself on the wrong side of the revolution. The National Fire Plan had been authorized at a moment of federal budget surplus. Now that the Bush administration had, as Vice President Cheney famously put it in another context, "other priorities," there was little interest in a possibly expensive program of ecological restoration in areas with few voters. Even biomass energy cannot compete with ethanol. A Forest Service audit on large fire suppression costs, which were skyrocketing, led the Office of Management and Budget to conclude that, with WFUs, the agency was "taking meaningful steps to address its management deficiencies," which is to say, working to get area burned up and fire costs down.[6]

But the deeper reason is that Wally's vision runs cross-grained to environmentalist enthusiasms. A small-wood harvest program might work but only if long-term contracts would allow access to that cellulose-clotted understory. That looks, or can be made to look, like logging by another name. The environmental community wants nature to restore fire; the fire community wants fire back, by whatever means works soonest and cheapest. The restorative agenda proposed by the Flagstaff model requires too much research, too much money, too much time, and it looks too much like silviculture by stealth.

As the era of big fires returned to the Kaibab, so too environmental groups had made their ambitions real. The Sierra Club and Center for Biological Diversity had shut down logging. The Grand Canyon Trust had bought out grazing rights. Collectively, they had established the northern goshawk as an index species of ponderosa pine health, made forest plans sensitive to the flammulated owl, and blocked off portions of the plateau as sanctuaries for the endangered Mexican spotted owl. Yet in one gulp the Warm fire had burned through reserved nesting sites for the owls and goshawk, wiped out a chunk of old growth, and shifted habitat away from pine-dependent species including the endemic Kaibab squirrel ("the owls and squirrel haven't found a way to live in aspen," Wally notes). A naïve observer might describe the outcome as the fire equivalent of a clear-cut.

Of course the Kaibab will not be destroyed by fire. Already greenery is poking through the charcoal. The scenic highway to the park will be awash with aspen, always a tourist delight. Mule deer will gorge on the sprouts. The biotic kaleidoscope will turn. The question is whether this is what we wish for the land and how we wish to achieve it. Wally belongs to an older school of conservation, a new-era forester read in Aldo Leopold, but still someone with a powerful sense of stewardship. People created the mess, they can't just walk away. If we value the goshawk, the flammulated owl, the tassel-eared squirrel, and old-growth ponderosa, then we have to intervene to save them. Eventually the Kaibab Plateau may fall largely under a regimen of naturally ignited fires; but the process of transformation is going to be awkward, and should be cautious, a voluntary rather than forced conversion. Wally worries that suddenly plastering the plateau with large burns will homogenize both the landscape and our options available for management—wonders whether the desired bold stroke, a series of shock-and-awe fires, will only catalyze an ecological insurgency that will be "unacceptable to future generations."

He voiced public concerns over the fire, and proposed to stage a workshop to examine the larger conservation issues of the Kaibab that the Warm fire's management raised. Forest Service officials then warned him that his public statements were becoming an issue. His criticisms might harm the cause of wildland fire use and "take away a tool from our toolbox" since the public, blinkered by Smokey Bear, couldn't be expected to understand the complexities of fire management. The important thing was to get fire back on the land, and the fire community had to stand united. He was also informed that funding looked bleak for his Ecological Restoration Institute. He was told there was no interest in a workshop on the Kaibab.[7]

Eventually the funding came through—those old political contacts paid off. But the workshop idea died stillborn. The North Kaibab conducted an "after action review" of the Warm fire that laid out the abundant flaws that led to the eruption and listed steps to prevent them from happening again. A more caustic draft asked "What Went Wrong???" and concluded: wrong place, wrong weather, wrong time of year, wrong fire-monitoring assessment, wrong results, wrong attitude, but conceded that in the end the "old standby, the weather" would get the blame. What was not at issue was the doctrine of WFU, which was fast becoming the treatment of choice for western wildlands. This was not the fire the North Kaibab had wanted; but it was a fire it was willing to accept as the cost of getting the burned acres it needed.[8]

Wally thought otherwise. "If you really want to destroy a ponderosa pine ecosystem," he argued, "graze the hell out of it, suppress fire, cut old-growth, and then let wildfire run amok." An "overzealous" WFU program could well be "the coup de grace" for the western woods.

Ultimately the issue was not about competing prescriptions but conflicting philosophies. Wally stood for an updated version of conservation in which humans had duties and had a responsibility to repair the damage they did and ensure that rare and valued natural assets got protection. This was, in a sense, a more ecologically sensitive version of the multiple-use doctrine that had guided much of Forest Service history. Removing people would remove all the good things people did. The more vocal environmentalists wanted a nature left alone, the sooner the better, and looked forward to a rewilding of public lands. Removing people would remove all the bad things people did. The assumption was that nature unaided would produce the best basket of environmental goods and services and that a naturally caused event like a lightning fire was nature's way of catalyzing the process. But wilderness is not identical with the natural, the historical, or the biodiverse. In the end it celebrates a transcendence of Nature, or what its proponents have always said it does, wildness. Wild fire may advance other environmental goals, or it may not. The only guaranteed outcome is a furtherance of the Wild.

Nor is fire simply a natural process or a "tool in the toolbox." It has its own logic, its own ancient alliance with humanity, and its own capacity for irony. It is not a mechanical implement like a fiery wood chipper, but a profoundly biological process that synthesizes its surroundings, all of it, including the legacies of human handling. It takes its character from its context. The Warm fire blew up because of conditions made possible by over a century of fire's exclusion. The downdrafts that drove the fire on its wild rush were the blowback from history. It remains to be seen whether the fire produces a biotic blowback of comparable dimensions.

That will depend on what already exists on the land and on what follows. The aspen rhizomes in the soil are suckering like wildflowers. The patches of exotic cheatgrass along the roadsides will propagate outward. Wildlife will search out new habitats, while owls and squirrels, and cougars and deer work out their complex choreographies. But all this is occurring amid a drought as serious as any the Southwest has experienced in a century (the 2002 season was the worst in a millennium), and amid the uncertainties of climate change. Allow wild fire to ramble and the only known you will produce is wild landscape.

And that of course is the crux. So powerful is fire that its management—and WFU is a form of management—can determine the larger uses of the land. The best way to control fire is to control its surroundings; but it is equally true that fires can define those surroundings. In the early years of colonial forestry, officials recognized that controlling fire was a means to control landscapes and the populations that used them. Deny fire, and you deny biotic access, and you can eventually change the land to something different. So, too, with fire's reintroduction. A program of WFU commits those lands to a wilderness ethos regardless of legal designation.

Once before the Kaibab had been, momentarily, at the vanguard of a national fire discourse. In 1890 John Wesley Powell barged into a meeting between Secretary of the Interior John Noble and the two leading American foresters of the day—Bernhard Fernow was the founding chief of the Bureau of Forestry, and Gifford Pinchot, the man who would succeed him and remake it into the Forest Service. This was a year before the country began creating forest reserves, and the two foresters were arguing, among other items, that such reserves were essential or else the land and timber would simply be lost to fire. Powell was then chief of the U.S. Geological Survey (USGS) and director of the Bureau of American Ethnology, which he founded, and was still widely honored as the man who made the first descent through the Grand Canyon in 1869. Powell's subsequent *Report on the Geographical and Geological Survey of the Rocky Mountain Region* had set up headquarters in Kanab, Utah, lying between the Kaibab and the High Plateaus, and commenced wide-ranging inquiries that led to the celebrated 1878 *Report on the Lands of the Arid Region of the United States*, one of the founding documents in American conservation. Among his conclusions was that fire took far more timber than the axe. The primary burning came from the hands of American Indians, and his crews codified that interpretation by mapping burned area in Utah. Remove the Indian, Powell admitted, and you would remove a major source of ignition. It was this Powell that Fernow and Pinchot expected would support their cause.

Instead Powell reversed himself and argued that the character of indigenous burning actually preserved the forest. Those fires were light, frequent, varied, and ineradicable; they kept the woods from exploding if left solely to the logic of nature. Contrary to forestry dictum, Powell urged a program that would emulate that local lore, what he had witnessed among the peoples he knew best, the Kaibab Paiute. Fernow and Pinchot were outraged. They dismissed Powell's perspective as an incitement to vandalism, rightly saw its challenge to forestry's axioms as a political threat, and denounced deliberate woodsburning as mere "Paiute forestry." These were, after all, peoples eking out meager existences and among the most technologically primitive in North America. That august professions like forestry, grounded in European academies, should recant their scientific sureties and accept the practices of peoples who lived on wild seed, lizards, and grasshoppers, and the occasional deer, was beneath consideration. When critics in the early days challenged the Forest Service doctrine on suppression, their calls for controlled or "light" burning were lumped with that old bugbear, Paiute forestry.

Wally of course saw the story differently, saw that the Kaibab had experienced over many millennia the presence of anthropogenic fire. The tribes had moved from lowland to plateau with the seasons, just as visitors do today. But to make the Kaibab habitable, they had burned, probably just as Powell and his colleagues described. Of course the Kaibab had ample lightning, and granted time enough, that lightning would by itself impose a regime on the land. But the regime that existed when sheep blunted the flames was not simply the outcome of natural ignition. It was a messy merger of torch and bolt. Likewise, many of those fires had supported hunting, which on the Kaibab meant mule deer. The saga of the Kaibab's disruption from historic conditions was not simply the result of removing cougars and wolves, or of suppressing lightning-kindled fires, but of also removing the keystone species for both: humans.

Wally read in that history a place for people. His critics did not, and reincarnated a version of Paiute forestry which this time held that people had been irrelevant or trivial. Whether they had burned or not did not matter since only the transcendent forces of nature such as climate could meaningfully shape the patterns of fire. Small numbers of wandering folk could no more alter those cosmic rhythms than could Kaibab squirrels. Their argument overlooks one of the defining features of fire, that it propagates, and it looks away from the inconvenient fact that it is humanity's combustion habits that are now shaping climate.

In the economy of nature, fire works with the "creative destruction" commonly attributed to freewheeling capitalism. What is happening throughout the West, either by intent or accident, is the ecological equivalent to the economic shock therapy urged upon the countries formerly under Soviet communism. Those committed to change argue that the sooner the transition, the better; the more quickly the invisible hand of nature can take over, the faster the necessary adjustment to a "natural" market economy. That the system may begin to experience wild booms and busts, or that some valuable features might be lost in the transition, or that mixed economies might work best, is irrelevant. What matters is to break the grip of the old system and allow the new order to begin. That is not Wally's way.

And that is why, at least in the short term, Wally Covington's strategy is unlikely to prevail. A beleaguered Bush administration was in no position to face down environmentalists, and its dismal budgetary standing left few options for anything that promised to demand federal expenditures, especially in areas with few voters. Even the fire community seemed to be abandoning patient restoration projects for more dramatic gestures. In May 2007 a fire officer and 30-year veteran of the Coconino National Forest around Flagstaff, Van Bateman, was convicted of arson for setting woods fires. His defense: he was only doing unofficially what onerous bureaucracy had made too cumbersome. He was getting fire back on the land, and insisted his behavior was "common practice." He might have added in the words of fireman Beatty in *Fahrenheit 451* that "fire was best for everything!" and that its "real beauty is that it destroys responsibility and consequences." The Forest Service of course denied Bateman's claim, but what shocked was not his justification but the cavalcade of defenders, many with honorable fire records, who rose to a kind of defense. Van Bateman was right: fire had to get back on the land. That this fire vigilantism was an eerie echo of the promiscuous woodsburning against which early fire protection had fought so bitterly seemed lost.

The scope for Wally's work may well shrink. There are too many critics who don't want to see the Flagstaff model expand beyond Flagstaff; who fear any return to an era of saw and torch; who doubt that anything people do can help nature; who find the agnosticism of letting nature choose too convenient a way to avoid humanity's moral agency. Wally's vision that his experiments might extend to the Kaibab and spare that beloved landscape from a violent conversion through a full-immersion baptism by fire is unlikely to happen.

Yet the work goes on. The Ecological Research Institute has students, projects, a slightly more secure funding; he remains a recognized presence, and for most, an honorable if misguided one (one of the "silverbacks," as he puts it). In the debate over the Kaibab he is marginalized rather than ostracized, an aging revolutionary who finds himself a member of the faction that accepted compromise when in revolutionary times it is the extremists who rise to power.

In a queasy way, he finds himself back in the cancer wards. He walks amid the charred yellow pine, shakes his head, yet wonders whether this might be a spot to maybe plant a handful of pine to replace those giants lost, to reestablish the conditions that prevailed when the forest seemed robust. Another spot he would leave alone. He remains a man who thinks in terms of things done on the ground, and amid the doubts there is also determination.

"I don't want to give up."

In 1968 lightning kindled a fire on the Dragon's Head, a pyramid-shaped butte in the middle of the Grand Canyon. Smoke drifted up, while beads of flame dribbled across the lightly vegetated limestone until they fell over the sheer wall of the Coconino sandstone and deeper into the gorge. Then the North Kaibab fire officer demanded that the park suppress that fire or else he would do it for them. The fire could go nowhere, and sending a B-17 to coat the Dragon's Head with diammonium phosphate was expensive; but that was not the point. The point was that suppression demanded instant action, and could not tolerate hesitation, slackers, or conscientious objectors. To let some fires burn out called into the question the premise that fire suppression was the basis for fire management, that the best way to prevent big fires was to control all fires at their start.

Thirty-nine years later, the Warm fire presented an unsettling symmetry. Its originating spark was not merely tolerated but promoted, and plans were altered opportunistically. The Dragon's Head fire could go nowhere; the Warm fire, everywhere. In the bad old days it was axiomatic that taking fire out could allow nature to recover from the abuses of settlement. That outcome proved true in many places; then it became an ideology, went too far, and the land became a mess. Today the postulate is that putting fire in will allow nature to recover, and that will likely prove true in many places. But how fire gets on the land will matter as will how we account for our oversight of that process. On such landscapes there is no obvious technical fix, as there is to prevent houses from burning.

The cycle of revolution was still turning, and was turning now on its own. Policy had changed, all for the good; the politics apparently had not. There could be no dissent from the proposition that fire was good and that more fire was better. The prophets and critics who had objected to suppression in the Sixties had bequeathed only acolytes, not replacements. They left no tradition of dissent. The old fundamentalisms had seemingly passed through a looking glass, reversing their image but leaving it otherwise intact. Fire's great cultural revolution appeared poised to replace one ruling elite with another, one defining ideology with another, one long-suffering landscape with another. Or perhaps not. Whether the outcome was the one desired was a matter of taste and cultural choice, a question of values. The one surety was that it would not be what everyone predicted.

The real issues are not about prescriptions for thinning or ignition patterns but about how we see ourselves in the world, about how a democracy reconciles its conflicting visions with a world of limestone and conifer and goshawk, about how it copes with uncertainty. Which is to say, it is a matter of politics. That is why Wally's voice needs to be heard, because the Kaibab is neither destroyed nor saved but in the process of radical change, and what needs to be preserved in fire management is that oft-frustrating, typically tedious process of democratic discussion, not only among squabbling factions of people but between them and nature. The radicals who stopped the agency from doing things they disliked want to shut off similar dissension now that it is doing what they wish. That, not the flames, is the concern, because what happens over the next several decades and amid a changing climate is no more likely to fit our forecasts than the FARSITE model predicted the Warm fire's blowup.

In 1936 the Civilian Conservation Corps constructed the original North Rim fire cache, a year after the adoption of the 10 a.m. policy by the Forest Service. When that policy began, the chief forester Gus Silcox declared that it was an "experiment on a continental scale" to stop abusive burning. It would happen, and then wither away as no longer necessary. Instead it endured, because what made it unsustainable was also what made it administratively attractive: its simplicity. It stated in unequivocal language what to do and how to measure success. The policy reforms that bubbled up from the Sixties had replaced that nominal simplicity with a charge to take "appropriate response." This prevented mindless adherence, but left unsettled just what action to take. Practitioners ached for some set of guidelines that would tell them what to do and how to judge their actions, for the uncertainties of fire are far vaster than its verities. A wildland fire use program promised to lift much of that burden from them. They didn't start the fire, and if it broke free, it was an inculpable act of nature. What finally condemned the 10 a.m. policy was not its determination to control fire but the administrative rigidity that hardened its vision into fundamentalism, and similar trends are the proper worry today.

The new cache stands ready for occupation. Perhaps 70 years from now some former rookie will return to marvel at its bulked-up size and modern apparatus, and then wonder how the fire landscapes of the Kaibab had come to change as they did and what those people were thinking when they made the decisions that let it happen.

Coda: Since my site visit, the 2008 election brought a change of administration that allowed a redefinition of appropriate (or strategic) responses that in turn ushered in another iteration in fire management. The wildland fire use fire joined the other "disappeared" of the fire revolution. What is replacing it is a doctrine of managed wildland fire that allows many responses to any particular fire and has quickly evolved into a strategy of box-and-burn, which is a variant of the long-permitted strategy of confining and containing. This is a strategy of active fireline engagement, not monitoring with suppression as an emergency backup. Part of what the new policy interpretation has boxed and burned is the seemingly unworkable practice of the wildland fire use fire.

SKY ISLANDS

ITS MOUNTAINS VARY by cause, contour, and significance. Some are volcanoes like the San Francisco Peaks, White Mountains, and Jemez. Many are tilted fault blocks, rising like reefs. Some are erosional, like the bold monadnocks of Monument Valley. The Kaibab is a dome. Shiprock a spire. Mount Graham a stony massif. But among that congerie the sky islands form a distinctive suite—Basin Range in origin, each isolated from the others but clustered collectively like an archipelago, all rising abruptly from a desert floor. They thrust upward, not grading gently from the lowlands any more than Tahiti grades upward from the Pacific.

They are small worlds unto themselves. Like other biogeographic isles, they are rife with endemism. In a region for which water means life, they are vital watersheds. Because they are wet, they grow combustibles; because they attract storms, they receive lightning; and because they are relatively self-contained, they suggest a unity of administration. They are the landscape unit that southwestern fire best recognizes.

RHYMES WITH CHIRICAHUA

THE CHIRICAHUA MOUNTAINS are famous for many reasons to many groups, but they are rarely known for their fires. They should be. Some start from lightning, some from ranchers. Some are set by rangers, or are allowed room to roam by them. Some are left by transients in the person of hunters, campers, and hikers. In recent years more are associated with traffic across the border with Mexico. The Chiricahuas have, at the moment, less of this than other border-hugging districts within the Coronado National Forest, though fires to distract, fires to hide, and fires abandoned by illegal border crossers are becoming more prominent.

Mark Twain once observed that history doesn't repeat itself but it sometimes rhymes. These days it seems there is a lot of rhyming in the Chiricahuas as fires echo a fabled but assumed-vanished past. This revival moves the Chiricahuas, among the most isolated of mountain ranges, a borderland setting for fire as for other matters, close to the core of contemporary thinking about managing fire in public wildlands.

The Chiricahuas—actually a giant, deeply eroded and flank-gouged massif—are among the southernmost of America's sky islands. They are famous for their powers of geographic concentration. Their rapid ascent creates in a few thousand vertical feet what, spread horizontally, would require a few thousand miles to replicate. Here, density replaces expansiveness. One can see across a hundred miles of sky, and into half a continent of ecosystems. It is possible to traverse from desert grassland to alpine Krummholz almost instantly.

They are equally renowned for their isolation, not only from the land surrounding them but from one another. The peaks array like stepping stones between the Sierra Madre Occidental and the Colorado Plateau; here, North America has pulled apart and the land has fallen between flanking subcontinental plateaus like a collapsed arch, leaving a jumble of basins and ranges as jagged mountains to poke through the rubble. The degree of geographic insularity is striking: they are mountain islands amid seas of desert and semiarid grasslands. On some peaks relict species survive from the Pleistocene; on others, new subspecies appear. No peak has everything the others do. A Nearctic biota mixes with a Neotropical one, black bear with jaguar, Steller's jay with thick-billed parrot. The Pinaleños have Engelmann spruce. Mount Graham boasts a red squirrel. The Pedragosas grow Apache pine. The Peloncillos are messy with overgrowth and dense litter; the Huachucas, breezy with oak savannas. The Madrean Archipelago displays the general with the distinct: unique variations amid a common climate. They can serve as a textbook example of island biogeography. That observation extends to their fires as well.

The Chiricahuas share this sense of the collective and the unique. From the distance, rising boldly, the Chiricahuas stand like a sentinel; up close, they act more as a portal. A portal for the onset of the Southwest's fire season. A portal for human traffic across the international border. A portal into a long-suffering discourse about the relative significance of lightning and people as sources of ignition, and hence as shapers of fire regimes.

It is a difficult geography to subsume either as an idea or an administrative entity. The last holdouts of the Apaches, for example, thrived by navigating its complicated terrain, including the politics of the international border with Mexico. (Geronimo finally surrendered at Skeleton Canyon within sight of the Chiricahuas.) Still, its isolation and its capacity to condense here achieve something that is difficult to do elsewhere. Sky islands can act as semicontrolled experiments, sieving out noise and distilling essences. So while the Chiricahuas might seem as remote from the national geography of fire as Easter Island, it highlights a theme of considerable significance in their curious display of fire's causation.

Throughout the Sky Islands fires roughly balance between human and natural causes; and this, historically, has been the case with the Chiricahuas. Lightning is abundant, and despite the rugged terrain—volcanic tuff eroded into deep gorges and pinnacles, a landscape topographically minced into an infinitude of fire behavior pixels—nature's ignition is more than ample to keep fire on the land. But ever since the climate stabilized after the Pleistocene, people have been on the scene, and they have burned. Both ignition sources, moreover, come embedded in larger geographic processes. Climate is a powerful presence that does more than kindle snags: it sculpts much of the overall physical geography. Likewise, people do more than throw sparks: they shape much of the landscape, particularly in ways that affect how a spark, once thrown, spreads or not. The proportions of fires set by each source—and the proportional contribution of each to fire regimes—rise and fall with the changing rhythms of climate and human migration.

There is little question that lightning is adequate to kindle copious fires and that the extent of burning aligns smartly with the ebb and flow of atmospheric moisture. Connect the sky island dots with the volcanic edge of the Colorado Plateau, and the resulting circle will trace the epicenter for lightning-caused fire in the United States. Like an outlying skerry that catches the first swells of an approaching storm, the bulky, border-hugging Chiricahuas make first contact with the Mexican monsoon, the signature onset of the Southwest's natural fire season.

This occurs annually. Regionally, there is a period of winter rains, followed by a long spring dry season, succeeded by summer thunderstorms as an inflow of moisture-laden air advances in a vast gyre from the Gulf of Mexico northward across the Mexican altiplano. The early storms, crackling with dry lightning, start the largest burns. As the rains continue, the land greens up, and although fires kindle in ever-greater numbers, they spread more weakly.

Atop this cycle of wetting and drying lie others, most prominently the El Niño Southern Oscillation that accounts for a peculiar cadence of multiyear wetting and drying. The ideal formula calls for several years of above-average moisture followed by drought. This pattern leaves more grass and shrubbery than organisms can crop off or decompose (in a semiarid climate like this the capacity for biological decay is scant). Fire takes the surplus.

But such observations are trivial. Of course fire obeys a logic by which wetting grows fuels and drying allows them to burn; that happens everywhere. Of course hot, dry, and windy conditions favor more fire than cold, wet, and calm ones; probably *Homo habilis* understood these dynamics. Documenting the relationship between fire's environment and its expansive presence would be embarrassing by itself, like noting that clear summer days are warmer than overcast ones. If this is the extent of documentation, then what Arthur Schlesinger Jr. once said of sociology, that it was the painful enumeration of the obvious, ought to apply to fire science's contribution to climatology.

But if the obvious beguiles, it is the second-order reasoning that proves treacherous. If you look at such data by itself, you might well conclude that climate alone "drives" the fire regime. Such analysis reduces a complex poker game to a game of solitaire: you can only play the cards nature hands you. The reality, however, is that there is another player at the table, and he is the dealer.

Humanity is the Earth's keystone species for fire, not only as a source of ignition but as a sculptor of landscape fuels. It is significant that this second source was present from the onset of the Holocene. There has been no time since the end of the last glacial when the region lacked an ignition source both more promiscuous and more prescribed than lightning.

From the creation, too, hominins have indirectly affected vegetation. They could do so by foraging, hunting, and generally fussing with the landscapes, usually with fire as a catalyst. The Southwest Pleistocene was a veritable Serengeti of megafauna, and fast-combusting fire had to compete with slow-combusting grazers and browsers. Then the megafauna—the mammoths, the Shasta ground sloths, the bison and glyptodonts—all disappeared. More and more it appears that the newly arrived humans were a catalyst in that vanishing act. They did not have to hunt every individual to extinction; they had only to add another predator to a crowded menagerie and to magnify the climatic impacts of warming through their own landscape burning. They shifted the fulcrum of climate. Much as an atlatl can add lethal leverage to a spear, so a favorable geography adds heft to human fire setting, and anthropogenic ignition expands the power of climate to affect landscapes. The elimination of megafauna liberated fine combustibles. The species that seized on the resulting surplus was humanity, which consumed it by fire, and through fire, reconstituted the landscape.[1]

For thousands of years, amid all the climatic wobbles that ended the Pleistocene, those fires followed prescriptions that fire history would characterize as aboriginal burning, which is to say, an alliance of torch and spear. Across almost all terrains and climates a common pattern emerges, as people brand the land with strips and patches, what might be termed lines of fire and fields of fire. The lines are corridors of travel; the fields, sites of recurring burning to assist hunting, to promote forage, and to harvest bulbs, grasses, nuts, honey, medicinal plants, and the like.

The resulting matrix is elastic. Tight terrain and a hostile climate might confine fires to distinctive blocks: they burn where they are lit with little outward spread, leaving a sharply etched mosaic. Where the landscape is open and rolling, the winds strong, and drought or dry spells frequent, the fires spread widely, and the mosaic becomes mobile not only in place but over time. Ignitions, moreover, move out of lower elevations into upper realms as flames spread under the influence of slopes as well as winds. In such circumstances the power of the torch can exceed the grasp of its handlers. And where people can amplify the amount of combustibles by removing faunal competitors, their power magnifies still further.

Nor do such fires behave only as people will them. Many result from carelessness or accident, or a kind of fire littering; and they obey the uncontrolled dictates of their surroundings. Some years they spread, some they don't. Some places get burned often, some rarely. Much as the removal of megafaunal competitors can cascade through a biota with unforeseen outcomes, so can the introduction of anthropogenic fire. Because the power of fire derives from the power to propagate, the source of human firepower resides, as so many aboriginal fire myths testify, in its setting. Nor are such fires always benign. Hunting fires can easily segue into fighting fires, as flame becomes, in contemporary language, weaponized. Hostile fire is as much a constant as fire hunts and smoke signals.

The Apaches exhibited a mix of calculated fire practices and fire littering. They put fire into the Chiricahuas through abandoned fires, signal fires, and fire hunting. They set patch burns for gardens and foraging. They set fires to encourage rain. They used smoke to lure fly-maddened deer. They kindled hostile fires. They set diversionary fires. They burned along their major corridors, which then, when conditions favored, moved up slopes and into protected niches, creeping and flaring as conditions warranted, not unlike the raiders who originally kindled them. And, as possible, the Apaches kept their active fires hidden from those seeking them. Renegade bands relocated out of the lower grasslands and into mountains, out of contact with formal patrols from presidio or cavalry post. But they did not, according to preserved records, kindle the kind of broadcast burns attributed to other aboriginal peoples. They didn't need to.

The dynamic of Chiricahuan fire regimes reflected this unstable interaction. When the rains came early and heavy, aboriginal fires fizzled out, and when drought followed a bout of wet years, fires blitzed well beyond campfire ring and fire drive. Overall, in places routinely visited by people, anthropogenic fires crowded out those of lightning; people burned first and preemptively seized the fire scene; they preferentially defined fire's regime.

No less than with climate, humanity's presence has swung in and out of cycles, in this case subject to the tidal and secular migrations of people and animals. The canonical Clovis site, where a distinctive lithic tradition met megafauna—a spearhead embedded in the bones of a mammoth—lies a little to the east. Cultures have been arriving and departing ever since. The collapse of the Anasazi and Hohokam civilizations in the 12th and 13th centuries created a major vacuum in the historical geography of human habitation. The Athabascan-speaking Apaches moved from the western grasslands into the grassy semidesert and then, at least seasonally, into the grassy-understory forests of the sky islands and Mogollon Rim. They carried their fire practices with them. The suppression of the Pueblo Revolt of 1680 created another rupture. The arrival of the Spanish mission system had consolidated some tribes while evicting and resettling others. In 1762, for example, Spanish officials effectively emptied the San Pedro Valley by relocating northern Pimas to the Santa Cruz River valley.[2]

That created a borderlands area between old-resident Pimas and new-arrival Apaches. The region underwent regime change from quasi-permanent habitation to intermittent occupation, as the landscape became a war zone, partly occupied, often fought over, burned for battles as well as for hunts and foraging. Then the historic dynamic reached far beyond the grasp of mission and presidio by replacing the extinct megafauna with horses, cattle, sheep, goats, burros, and swine, all of which competed not only with wildlife but with fire. In principle, the combustibles that the late Pleistocene extinctions had liberated, the late Anthropocene would again corral, sending them into the gullets of livestock. In practice, that colonization first required the pacification and relocation of the Apaches. Not until the indigenes were suppressed did livestock successfully overrun the ranges. Eerily, the year of Geronimo's surrender, 1886, saw the last great breakout of fire in the Chiricahuas. Free-burning flame had less to feed on, and it starved. Even before the formal policies of suppression, the mountains had entered an era of fire famine.[3]

Of course the movement of imperial people into and out of the region—what anthropologist Edward Spicer in a larger context famously described as "cycles of conquest"—had to interact with the cycles of climate to yield the region's chronicle of fire. And that is precisely the point: it is the *interaction* of these two grand rhythms, one of wetting and dry, and one of human coming and going, that the region's fire records testify to. There is little dispute that massive overgrazing beginning in the 1880s coincided with a monumental drought in 1891 to drive the high grasslands and forest savannas into collapse; and by destroying surface combustibles, this one-two punch knocked fire out of the biota for decades. It was, however, the removal of the Apache that allowed for the wholesale reintroduction of grazers. The extant fire regime received a triple blow: one from climate, one from fuels, and one from ignition sources. Two of the three were the outcome of people.[4]

On all these counts controversy has flared, and it typically pivots on how much agency to grant humans. The critics claimed the hard, high ground of science, dismissing outright appeals for human agency or herding them into disciplinary reservations where they feed on the lean rations of anecdote. Climate change, not spears and torches, must have driven megafauna over an evolutionary cliff. Climate change, not longhorns and shovels, must be responsible for squeezing fire out of mountain and grassland. Behind the conviction lay an insistence that one or the other cause must dominate.

The motives behind this reasoning are not difficult to discern. They are, first, often concerned less about the past than the future. If the protected sites are not "natural" but cultural landscapes, then the passage is open for people to reintroduce not merely fire but hunting, and then grazing, and ultimately to follow a slippery logic that must lead to trailer parks and casinos and "wisdom sitting" on slot machines. But the outcome can also challenge science because it says the numbers generated from tree ring scars are not simple "proxies" of climate but indices of people and climate interacting in complex ways. That muddies not only the chronologies but the epistemological status of fire science, not to mention its funding. And not least, perhaps, is that old yearning for an Unmoved Mover. If people have shaped everything, there is no escape from our postmodern selves. The landscape becomes a Möbius strip. Granting agency to a few lightning bolts seems a small price to pay to keep Nature's God, however secularized, in His heavens.

Yet a simpler explanation may also be at work, which points to timing, a coincidence not of climate but of culture. It is an accident of history that formal scholarship came to bear on the topic during a period when the human presence as a fire lighter had been stripped away and human agency as a fire fighter had became prominent. The debate about the relative power of nature and culture centered on lands deliberately emptied of most human activities, and for which almost all human fire practices were banned, save the relentless task of suppressing whatever fires might start. It was a kind of reverse reservation system, one intended not to keep people in but to keep them out.

This coincidence fundamentally distorted an emerging discourse. When observers—natural scientists—inquired into the causes of stream trenching, creeping desertification, and the smothering propagation of woody weeds, they did so not only with the instruments of their disciplines, but during a cycle of human migration in which anthropogenic fire had bottomed into a deep trough and enthusiasm for restoring "natural" fire was entering a crest. In the early days of state-sponsored conservation, right thinkers aligned fires started by people with overhunting, abandoned clear-cuts, and livestock that ate and trampled everything in their path, or what Teddy Roosevelt called "scalping" the land. People frequently burned promiscuously; lightning fires seemed lost in the mix. Then cultural interests shifted toward a fascination for the pure wild; for preserving unique habitats, for reintroducing wolves, for restoring lightning fire. Anthropogenic fire was deemed unnecessary and intrusive, and if used it was justified, rather on the logic of *Star Trek*'s Prime Directive, to correct the errors of past intervention. Natural fire was by itself necessary, sufficient, and inevitable.

This discourse—there was not enough resistance to warrant calling it a debate—was both skewed and curiously scholastic. It ignored, for instance, the most fundamental of the facts before it: that the vacated "wildlands" existed because of cultural decisions. These were wildlands free to express natural causes because people had chosen to make them so, because they had transcribed political values onto raw geography. Had public lands not existed, or had they not been moving toward a wilderness model of management, much of the debate would have been even more fatuous. And by restricting itself to certain kinds of evidence, the participants lacked the power to resolve their evolving discussion. Instead, the reasoning suddenly entrenched itself, like the San Pedro River, and any new water flow had to follow that deepened discourse.

The historic records are as dispersed and eccentric as the sky islands, and they are intrinsically suspect to those who demand clear signatures on natural archives like tree rings capable of quantification. For cultural archives, what gets recorded depends more on who does (or does not) do the recording than whether there is anything to record. Most early European observers were military patrols or missionaries, and the Apaches who flourished in the mountains had little interest in either. It was in the Chiricahuas particularly that they made their last defiant stand, using the international border and offbeat mountain trails to frustrate efforts to intercept them. Naturalists (and anthropologists) arrived much later, after the old regime had become scrambled; for the most part, after the cultural landscapes of the Apache had been suppressed, then sequestered, and finally dissolved. The only fires that persisted came from lightning since all lightning's competitors were gone.

What happened to the people happened also to their documentation: they got shoved into reservations outside the mainstream of fire science. Researchers examined them as they would dendrochronology, as packets of data, oblivious to their context or character. Henry Dobyns has wonderfully described, for example, the ability of the Apaches to avoid being seen by those looking for them. Since any smoke would advertise their position, they shunned fires when hostile observers were in the region. And until Apache scouts were exploited for counterinsurgency, it was easy for renegade bands to track and avoid the movements of military patrols whose journals have been a major source of ethnographic and ecological information. He observes for the Mormon Battalion, which included "a number of wagons," that the operation "engaged in the highly conspicuous activity of opening a wagon road through the Apacheria. A unit more unlikely to see hostile Apaches would be difficult to imagine." Of later patrols by dragoons officered by "a more or less constantly intoxicated commander," he comments wryly, that the "Apaches might well have had to attack this command in order to gain its attention."[5]

It's a classic case of the absence of evidence not indicating evidence of absence. In fact, there are more eyewitness records of anthropogenic ignitions than of lightning. If one demanded the same standard for both fire causes, we would have to dismiss natural fires as trivial or an outright fabrication.

The fires reached a low in the 1920s. Too much had happened, and unlike the revolution that had ended the Pleistocene, which had leveraged the regional fire load upward, these reforms worked to depress it. Fire prevention was an agency goal, fire fighting had become more effective, forests were felled, woodlands fed into charcoal, and grazing, while reduced, was more than the mauled land could handle. But the fires were also receding from the rural countryside generally, as industrialization offered flame-free substitutes, bolstered suppression with trucks and pumps, and redirected the regional economy.

Everything worked to further dampen fire. The Depression knocked away what props remained to commodity production. The government stimulus by the Roosevelt administration that replaced it unleashed, among other programs, the Civilian Conservation Corps, which led to a flurry of fire roads, lookout towers, and crews. Lightning and accident still kindled blazes in the hundreds; but they had little chance to roam before being attacked and they had scant forage to feed upon, even as woody flora began replacing grasses and forbs. By the 1950s, with war-surplus jeeps, trucks, and planes ordered to firelines, the landscape had changed irrevocably. It could no longer support the old fire regimes: consistent anthropogenic fires were gone and the predominantly grassy fuels that had carried flames like gentle winds ruffling muhly and fescue had been forced aside in favor of dog-hair thickets of pine, scrubby understories, brushy hillsides, and woody litter thrown over whole mountains like a vast eiderdown comforter. Even as the mid-1950s experienced the worst drought in decades, burned area plunged to token fractions of its former extent.

Research capabilities evolved, and explained the overturn of the biota by appeal to climatic cycles, galvanized by that late 19th-century irruption of livestock. Still, the cattle and sheep were regarded as a one-off event: it was the implacable rhythms of climate that laid down the fundamental reality. Certainly, this was the conclusion to their 1965 classic, *The Changing Mile*, in which James Hastings of the University of Arizona and Raymond Turner of the U.S. Geological Survey exploited repeat photography to record a massive shift from grasses to desert succulents and woody biotas particularly within the San Pedro watershed. They dismissed as incompetent "the fire hypothesis," which implicated fire's extirpation as a prominent factor. The more powerful institution was the Laboratory of Tree-Ring Research, first established in 1937 by the astronomer A. E. Douglass to record the effect of the sunspot cycle on climate. Dendrochronology easily segued from measuring the width of tree rings as an indication of climatic fluctuations into chronicling the abundant fire scars branded onto them. Fire scars thus became a proxy for climate. The emerging annals confirmed what seemed self-evident to anyone who had studied the region over the past few decades: climate was the primary fact and force of change. And since lightning was an expression of climate, the primordial order of fire followed the syncopation of spark, rain, and drought.[6]

Fire was natural—that much was obvious. What made it natural was its origin in climate, and if fire's presence had diminished as an ecological enterprise over the past century, that was because people had unwisely meddled in nature's economy. Those interventions could not endure. Irruptions of cattle and experiments in fire exclusion would pass. Ultimately, climate would reassert its implacable logic.

By the time *The Changing Mile* saw print, America was fast spiraling toward its great cultural revolution on fire. The origins of that reformation are several, but for the public wildlands, the irrefutable argument was, fires are natural, and parks and wilderness areas ought to promote them as they would other expressions of the Wild. At the minimum they ought to stop trying to suppress them, which had in any event failed and had wrecked far worst havoc on the regional landscape than wild fire ever did. The war on fire had failed: it had only stirred up an ecological insurgency that no summer surges of firefighters and air tankers could contain. By 1968 the National Park Service had officially reformed its policy in an effort to promote more burning; a decade later, the U.S. Forest Service followed. By 1995 all the federal agencies had a common policy, reemphasized in 2001.

Actual change on the ground was slow, as ideas proved easier to apply in seminars and conference rooms than amid pine thickets and manzanita. But the direction of movement was clear, and research consistently pointed to the fire-climate bond as the primary driver of fire's ecological presence, as an argument to promote more burning, and as the standard for what restoration might achieve. As large patches of land—Saguaro National Monument in the Rincon Mountains, the Gila Wilderness in the Mogollon Mountains—allowed more room for lightning-kindled fire, something of a long chronicle of jostling burns emerged as empirical evidence that nature, and nature alone, could establish fire regimes, and that nature alone ought to do so. The mechanisms were doctrines of prescribed natural fire, later renamed wildland fire use, and bestowed with other euphemisms that disguise what ordinary people would call let-burning.

There were practical reasons for outsourcing the task to nature, principally safety and cost, for such fires proved much cheaper than the alternatives, at least until a fire broke its fetters, at which point it became more expensive and damaging to releash. But the primary reason was ideological. Although deliberate prescribed burning had become an acceptable practice, it was generally regarded as a best-of-evils surrogate. It was a transitional phenomenon that would fade away as nature, under the implacable impress of climate, reclaimed more and more of its former dominion. Human ingenuity, arrogance, and mechanical might could not resist such indomitable forces as the El Niño Southern Oscillation or the Atlantic Multidecadal Oscillation that, in the Southwest, chose to express themselves as fire. People could not exclude fire because they could not control climate.

This was for decades the authorized version of how and why fire has reasserted itself. People have backed down from their idiotic and impossible attempts to stem the climatic tides, and have begun some measure of atonement by deliberately reinstating fire to amend for the years lost. Given time, climate will purge away the contaminants introduced by ranching and firefighting. A purer Nature will reclaim the landscape, under the distant direction of that Unmoved Mover, climate.

The contemporary Chiricahuas, however, suggest another narrative. To the impossibility (and undesirability) of excluding lightning fire they add the impossibility (and undesirability) of excluding anthropogenic fire. The core issue is not whether fire is present or not but fire's regime, and for this people as nature's keystone species over ignition must return along with lightning. What is interesting about the Chiricahuas is that each is active, both as an expression of official policy and quite in defiance of it. These fires can no more be stopped without ecological unrest than can lightning's.

Rangers and ranchers are reintroducing prescribed fire. A consortium, the Malpai Borderlands Group (MBG), has created an institutional matrix for coordinating burns among the Forest Service, National Park Service, Bureau of Land Management, Arizona State Forestry Division, the Nature Conservancy, and private landowners. The Feds do it as a way to supplement natural ignition, a kind of performance enhancer. The ranchers do it because fire is the most powerful and cheapest way to stimulate the complex forage on which their way of life depends. The ancient rhythms of fire hunting have returned in the avatar of fire herding.

But the most interesting reintroduction is occurring outside official channels altogether. The Chiricahuas and neighboring Peloncillos have reestablished themselves as a portal for unauthorized human traffic across the international border—a veritable Mexican monsoon of border crossers who carry fire as much as contraband. The variety and geography of the burns eerily echo the old Apache suite: abandoned cooking and warming fires, accidental fires, fires set as distress signals, fires kindled to divert attention away from illegal activities. The border patrol has proved no more effective in stemming such ignitions than the Forest Service was, over the long run, with lightning. In the end, both have proved unable to shut down the fires, and perhaps are unwilling to do so at the costs demanded. Officials can't turn off lightning, and they can't control people who are by definition renegades and "illegals." Fire has returned. The restored rhythms are restoring the old rhymes.

The mountains are frustrating efforts to assert the primacy of one or the other ignition source. Both are ample, and in such intricately dissected terrain, washed over by cycles of climate and human migration, neither dominates. The insistence that in the name of good science or adequate nature protection sky-island fire must be natural is fading in favor of a belief that fire will come—from many sources—and that it will be better if humans have some rational say in how it comes than to accept whatever accidents history serves up. Intellectual borders are being overwhelmed by the reality of what is happening on the ground.

In this revised discourse, the Chiricahuas may be synecdoche for the story of how we understand fire nationally. The right fire regimes will mix both ignition sources and landscape sculptings. A proper scholarship will begin with the axiom that they interact, not that one or the other must be the designated driver. In truth, it makes more sense to imagine fire, in particular, as a driverless car integrating all the factors around it as it barrels down the road.

What should matter is how fires, of any and all kinds, play out on the land around them. There will be a place for lightning-kindled blazes. And there will be a place for anthropogenic burning. The question that has tyrannized fire management since the 1960s—whether a fire is natural or not—hardly matters. So too the old obsession of fire science—whether nature or culture is the true irresistible force—is revealed for what it really is, a metaphysical query, a proxy not for "climate" but for the values of climate researchers. One might as well ask how many lightning fires can dance on the summit of Dos Cabezas.

The pace of reform is quickening. Nationally, wildland fire use is being discarded, as the prescribed natural fire was before it, and as the "resource benefits" burn will be in the future. Under a doctrine of appropriate management response, there is only fire. Natural fire need not, as an initial response, be controlled at the smallest acreage. A wildfire might be held at some natural barrier, or tolerated, or even encouraged. In 2002 and 2003 the Aspen and Bulloch fires together burned 46,000 acres, or roughly 85 percent of the Santa Catalinas. In 2005 the Coronado National Forest amended its land management plan to allow for natural fires outside legally designated wilderness. In 2007 the Coronado began to scale up that new strategy; in 2008 it racked up 10,000 burned acres under its aegis; in early 2009, amid "unusual" weather, it approached nearly 20,000 burned acres even prior to the onset of its traditional fire season. In 2011 the Monument and Murphy fires swept over almost 100,000 acres, while the Horseshoe II fire rambled across 223,000 acres in the Chiricahuas. Collectively, the 2011 fires burned one-sixth of the sky islands under national forest. The characteristic fire of the Sky Islands was evolving into a box-and-burn hybrid of the wild and the controlled that seems particularly apt for a region so long characterized by a fusion of nature and culture.

But that reclassification also means there is no intrinsic reason to swat out anthropogenic ignitions either. The old dichotomy among fire sources—natural and human, deliberate and accidental, malicious and benevolent—dissolves, as it becomes more difficult to police the policy borders that have discriminated among practices according to their sources. In truth the intellectual case is even more untenable as people have shifted their fire practices to industrial combustion—as they have replaced the open burning of biomass with the machine combustion of fossil fuels. This is proving an oscillation more powerful for pyrogeography than El Niño since it is evidently unhinging not only terrestrial biotas but the Earth's climate itself, including the rhythms of El Niño. Revealingly, half the Southwest's megafires in the new millennium have been started by people and half by lightning.

Such is the reality of fire in the Chiricahuas. For a long time it was not recognized by fire science, which achieved much of its clarity by excising people just as it has omitted their fires. Take away human agency and only natural causes remain, and with only natural causes, the sole medium of research must fall to natural science. The rest of the syllogism follows. Only by knowing the proper mechanisms—linked in a chain of causality—can we devise suitable responses; only natural science can track that succession of attribution; and only such sciences can suitably translate the explicated chain of consequences into a chain of command by which research informs and management applies. But times change, and so does science along with climate and landscapes and human demographics. The overwhelming human presence in today's world is forcing a reconsideration of the past.

It was Aristotle who observed that without an unmovable final cause explanations would slide into an infinite regress. In order to succeed, an explanation, much as with a narrative, needs a fixed end. Otherwise each cause only leads to another, like a Sisyphean scavenger hunt. For fire science that final fixed mover—the prime mover unmoved—has been climate. But as anthropogenic combustion habits destabilize the atmosphere, the climate is no longer an Unmoved Mover (or an alien Other), but an extension of humanity's confused agency. The assurance that climate is the driver blurs. It may be only a question of time before such realizations cross the border that divides the two cultures of science and humanities in such numbers that they force fire scholarship to accommodate people as it has nature. That powerful arguments are being mustered to identify the contemporary period as the Anthropocene suggests a change in the climate of opinion is underway.

For now, Chiricahuan fires are repeating a historical refrain. It may be that over the coming decades either their yin or their yang will once again go out of phase, that one or the other ignition source will be actively encouraged or deliberately suppressed. If so, one might expect, a few decades after such a misguided experiment, that the old refrain will once more reassert itself, and we will discover yet again that, however linguistically awkward, fire continues to rhyme with Chiricahua.

TOP-DOWN ECOLOGY

Mount Graham

A LONG NARROW ROAD winds steeply up into thickly wooded backcountry to an exclusive enclave of costly structures, all well beyond the periphery of settlement. It's the formula for the worst-case scenario of the wildland-urban interface, except that this is no subprime landscape stuffed with trophy homes. It's a telescope complex atop Mount Graham, and on the sky islands of Arizona the scene is repeated four times. Call it the wildland-science interface.[1]

Fire management accepts as axiomatic that it is science based or at least science informed and that good science is the antidote to the toxins of politics, land development, and a Smokey-blinkered populace that doesn't understand the natural ecology and inevitability of fire. Science is better than experience or history, and more science is better still. Science, preferably natural science, since even social science is tainted with the implied values of its human doers, is the solution. At Mount Graham, however, science is the problem. And the challenge is not simply that "science" here underwrites its own version of the WUI and opens paved roads to remote sites that complicate fire management and compromise biodiversity. The real challenge is the assumption that science stands apart from the scene it describes and from its Olympian heights can peer objectively outward and advise wisely.

The Mount Graham International Observatory (MGIO) suggests instead that science's lofty perch is not removed from land management and that science, too, has its self-interests that can influence what it sees, does, and says. Science, in brief, is not an ungrounded platform for viewing the universe of fire and recording its observations. It is sited, and that siting determines what it sees, and decisions over such sites make science and its caste of practitioners as motivated by their own values and ambitions as loggers, ranchers, real estate developers, and ATV recreationists. Science has its own dynamic apart from nature, its own presence on the land, and its own politics. The 1.83-meter primary mirror of the Vatican Advanced Technology Telescope, while nominally looking out, is also a reflecting lens that looks back on its viewers.

As their name hints, the sky islands are ideal for astronomical observatories. They sit atop high mountains amid a dry climate surrounded by dark deserts (the exception is Tucson, but the city has adopted light-abatement measures). The costs of constructing and operating such facilities favor clustering, and the region is dense with telescopes. Mount Lemmon in the Santa Catalinas holds one for the Steward Observatory, Mount Hopkins in the Santa Ritas hosts the Whipple Observatory's MMT for the Smithsonian Institution Astrophysical Observatory, Kitt Peak in the Baboquivaris has a compound of 22 instruments including the famous National Solar Observatory, and in the Pinaleños three telescopes sit in a concrete aerie atop Mount Graham. In recent years wildfires have threatened them all.

The fires have come almost annually. In 2002 and 2003 the Bulloch and Aspen fires together burned 85 percent of the Santa Catalinas. In 2004 the Nuttall fire complex burned 29,000 acres of Mount Graham. In 2005 the Florida Peak fire in the Santa Ritas threatened both the MMT telescope and cabin inholdings in Madera Canyon, and forced evacuations. In 2007 the Alambre fire moved up the slopes of Kitt Peak before being contained at 7,000 acres and over $2 million in suppression costs. But such scenes are hardly news: the same dynamic is playing out across the country, and for that matter, throughout the industrial world, as a revanchist vegetation meets an out-migration of urbanites. Matter and antimatter—the astrophysicists needn't peer into nebulae at the fringes of an expanding universe to detect such explosions. They need only look around them.

But the deeper collision is occurring within the domain of cultural values, not subatomic particles. The observatories break up public wildlands into incommensurable blocks: they are in this respect no different from a private inholding or a clear-cut. Only four sky islands have roads that extend to their summits; all lead to observatories. So long as science seemed a greater good, as incontestable in its claims to public land as to public money, there were few objections. Certainly there were no doubts from the scientific or university communities. They were the good guys, far removed from grubby commodity producers and selfish summer homeowners. Their motives were unimpeachable. They were studying the heavens. Theirs were the highest values of civilization.

Then the science-industrial complex met the Wilderness Act, the American Indian Religious Freedom Act, and the Endangered Species Act. They spiraled together with particular force at Mount Graham when in 1984 the University of Arizona (UofA), heading a consortium, petitioned to create a cluster of seven telescopes, one of them an enormous six-dish, rail-mounted interferometer, at the summit. That catalyzed an opposition. An Apache Survival Coalition declared the peak a sacred mountain. Advocates for roadless areas wanted access limited rather than enlarged. And enthusiasts for wilderness and biodiversity noted that the mountain was a Pleistocene relic of Englemann spruce and corkbark fir with 17 protected species, including a unique subspecies of red squirrel that inhabited the summit, and could go nowhere else; expanding the facility over two peaks would diminish its required habitat and perhaps introduce other disturbances. Under terms of the Endangered Species Act environmental groups protested and eventually brought suit.

The controversy—"scopes vs. squirrels"—became more bitter as the years passed, not lessened by the inability of the science community to admit that they were in fact upsetting a biotic order. Compared to the 156-billion-light-year width of a Hubble universe, Earth is less than a fly-speck and the addition of a few acres of telescopes on the Pinaleños tiny beyond infinitesimal. But compared to the habitat of the red squirrel and public claims on a patch of land that could not enlarge or go elsewhere, it was a significant, probably irreversible disturbance. The University of Arizona and the astronomical community refused to accept that fact or to place themselves within the scene. The Mount Graham International Observatory was only a platform for viewing. MGIO was not itself within the panorama viewed.

The controversy dragged on for years, a bitter war of bureaucratic attrition. On-site protests, fudged reports on the biological status of the squirrel in particular, intervention by the state's politicians, backroom deals, legal suits, court injunctions, appeals—from the time the UofA proposed a complex akin to that on Kitt Peak until three telescopes actually arose on Emerald Peak and an adjacent knob, a decade of rancor passed. Throughout, the tendency was to interpret the feud along familiar tropes such as jobs vs. environment, or the perversion of biological science by politics, greed, and hubris. For environmentalists, Mount Graham joined Glen Canyon as a martyred landscape. Yet the contest might as equally be viewed as one between sciences, and between science and wildland management, and between institutions of science. One science, astronomy, and a nominally science-supporting institution, the UofA, turned to politics to overturn the claims of another science and its nongovernmental auxiliary. Astronomy meant big science, while conservation biology had only acquired a name in 1978. Deep sky met deep biology, and sky won.[2]

In retrospect, it should have been obvious that Mount Graham would become a point of convergence for controversy, as much a focal point for gathering environmental themes as the lenses of the observatory were for light and microwave radiation in the night sky. That applies not only to its ecological status but to its human history.

The mountain is like a living natural-history cabinet of southwestern species. The Chihuahuan Desert laps against its eastern flank; the Sonoran Desert, its western. The floodplain of the Gila River runs along its north, and high desert along its south. So steep is the mountain that life zones become wafer thin, stacked like poker chips and tucked away into niches amid the deeply crenulated flanks. From Fort Grant on its southern shoulder to the peak of Mount Graham six horizontal miles rise 6,000 feet, or better than one foot of rise for each six feet of run. Forest types appear like thin sections from rocks; only along the rough-rolling summit is there breadth enough to create the semblance of a forest, and those 3,000-plus acres of Engelmann spruce and corkbark fir are a relic biota, a Pleistocene refugee as distinct as the California condor, and just as endangered.

The mountain's human history has been as mixed. Major mines bore into the mountains to the north. Ranching overran the lowlands as soon as mines developed and the indigenous tribes were pacified. Mormon settlers colonized the Gila River floodplain for irrigation agriculture, establishing regional entrepôts like Safford and Thatcher. Military posts dot the region; Fort Grant was part of General George Crook's famous campaign to contain the Apaches, blocking potential escape routes from the San Carlos Reservation to Mexico. The Mormons and the military, from the north and south respectively, converged on the summit. Settlers developed flume-transport logging that ate away at the northern canyons. The army established a hill station, complete with a hospital and heliograph lookout, along the summit, with a wagon road to supply it. The settlers deposited a small cluster of cabins at Columbine, while Fort Grant evolved into a state prison, which continues to supply labor for forest-related projects. In 1902 Mount Graham became a forest reserve; and after the U.S. Forest Service acquired control over the reserves in 1905, a national forest in 1907. The federal government went from suppressing Apaches to suppressing fire; Heliograph Peak became a fire lookout; fires all but disappeared. Although formally vanquished, the Apaches continued to identify sacred springs like the Bear Wallow cienega on the top, and ritually revisit them. Then the University of Arizona and big-science astronomy staked a claim.

By that time Mount Graham, despite its extraordinary ruggedness, was becoming an ecological shambles. Logging, grazing, fire control, the introduction of the Abert's squirrel by the Arizona Game and Fish Department (as a meat source), all had broken the biological integrity of the mountain refugia and compromised its resilience. In addition to its summit woods, as isolated as though they were on Selkirk Island, the mountain had the northern goshawk, the Mexican spotted owl, and the red squirrel, all threatened or endangered. But underwriting everything was a thickening bloat of combustibles, both choking the old biome and stoking the potential for large fires. The scene worsened when spruce beetles (native) and spruce aphids (exotic) began stripping spruce as part of the living dynamic. The tiles of the old mosaic began falling away.

Some of this was obvious when the UofA declared its intentions. The potential biological (and cultural) competition was clear from the onset. What all sides failed to consider, however, was fire. It claimed no more than a nominal paragraph in draft assessments, as bureaucratically trivial as a burrowing mammal.

One reason is that attention was riveted on the summit and its residual woods, widely regarded as an asbestos forest. It had never burned in the memory of American settlement. Later studies demonstrated that the peaks had not burned for 300 years. (In a gesture of historical irony, the last major fire on the heights of Mount Graham had occurred in 1685, two years before Isaac Newton published the *Principia Mathematica*, the foundational opus for modern astronomy.) Lightning kindled no more than 6 to 12 fires a year, most of which quickly self-extinguished. People had long ceased to set burns deliberately. Fire seemed a relic from the past, like the heliographs now replaced by microwave repeaters.

The critical concern with fire, moreover, did not reside at the summit but along the lower elevations. Few ignitions would start at the top: there, lightning would typically be accompanied by copious rain, and spruce fir lack the long needles that make an ideal bed for surface burning. Instead, ignitions would blast up the slopes from lower elevations; and part of what had spared the summits from repeated fires was that those montane woodlands had absorbed routine burns without blowing up and hurling flames to the crestline. That middle landscape was the one most in upheaval: it was fast morphing from a predominantly open Douglas fir woods to a mixed forest choking with assorted conifers, all as congested as a squirrel midden and as combustible as a crate of excelsior. Much of the change had occurred over the past 60 years. Now, if drought drained those fuels, fires could burn unheedingly across the old borders. The interface between woods had blurred and the interface between woods and scopes had sharpened.

Certainly, this was the reasoning of fire behavior science, which now also found itself on the summit, overlooking a fireshed much as the Large Binocular Telescope does the Milky Way. Here it appears to challenge the other sciences for space, and in fact might even seem to synthesize them, as it studies the bio-burning of the landscape by the methods of physics. The logic of fire behavior would appear to favor its claims. Fires burn most fiercely where they have more to burn, which at Mount Graham is also where ignition is most common. Fires burn most savagely upslope, and both MGIO and the squirrel's habitat are at the top of the mountain. Add in a warming climate, in which the desert seas will rise, and fire may remake the landscape as thoroughly as glacial ice in the Pleistocene. In time, without aggressive action—not just suppression but preventative intervention—fire might well claim it all, like a sun going into supernova and taking its planets with it.

Then life imitated science. In May 1996 the Clark Peak fire, kindled from an abandoned campfire, burned 6,500 acres, spreading into known squirrel middens, giving the MGIO a thorough scare, and inspiring a fire inspection by a scientist who specialized in the WUI. In June 2004 lightning sparked two fires, the Gibson and the Nuttall, that together drove through nearly 29,000 acres on the mountain's northeast slopes and nearly converged exactly at the MGIO. All three fires emerged from the mountain's middle zone. In the end, both squirrels and scopes survived. The squirrels took the bigger hit since the beetles and aphids had struck hard even before the burns. Afterwards, the MGIO hardened its structures to prevent embers from entering the interior and burning the scopes from the inside out, and it laid out perimeter protection in the form of a network of sprinklers. Besides, bugs and burns had now insulated it from the threat in ways not possible for the squirrels. The observatory did not live off the lost trees.

It would seem that fire science also exerts a claim to the summit. Yet since science consists of methods and ideas, not a physical entity, how might it demand space akin to that of the Heinrich Heitz Submillimeter Telescope or the Mount Graham red squirrel? In fact, it can because it affects fire's management on the ground. Here is another wildland-science interface, where science as a mode of inquiry confronts fire management's need to act, and of the two wildland borders, this is much the trickier.

There are those who say that the barrier should not—does not—exist, that fire management is simply the best application of the best knowledge science produces, and that more science will spark better practices. But there are also those—fire's curmudgeons—who point out that humanity handled free-burning fire, and probably handled it far longer and more successfully, before modern science than after it, and that more money poured into research has not produced savings in fire management's costs or a reduction in area burned by wildfire. Rather than improve performance, science as an intellectual enterprise sits within fire management as the MGIO does on the peaks of the Pinaleños. It not only interrupts the landscape of fire but demands a peak and shoves aside competing claims. What then should be the relationship between knowing and doing, and what is the proper place of science?

From its origins in forestry, fire protection, and later fire management, have insisted that they are science based. It has been their self-declared mission to counter centuries of misguided folklore and superstition, and they, and they alone, should instruct administrative practice. Yet the record of their accomplishments is at best mixed. Science proved excellent at reducing complex systems into simple matrices and then creating machines (or rules) to apply particular actions. Fire protection became an intellectual complement to silviculture: it could successfully decrease burned area much as tree farmers could pump up cellulose production. But public wildlands are not pine plantations.

The outcome was generally disastrous. Forestry used its status as an academic science to counter folk wisdom, to condemn western underburning as Paiute forestry and cracker-cowboy range burning as wanton vandalism, and to dismiss counterclaims by prairie restorationists and wildlife biologists as the wistful fancies of niche hobbyists. It used science to condemn fire's presence on the land and denounce its practitioners. When experiments in the southern pines showed favorable results from burning, the forestry profession and its state-sponsored expression, the U.S. Forest Service, suppressed them with the same fervor it attacked flames. Eventually, the outsiders were able to replace the "bad science" with their "good science," but the environmental damage had been done.

The response? An ardent appeal for more and better science. If the agencies had had more of the right science, so the apologists claim, they would have applied those findings and could have resisted the perverting pressures of politics. Rather, the story seems to be that the agency did have the best science of its day and applied it with consequences later scientists rejected. Still, one could legitimately discriminate between science as a mode of inquiry and science as a body of positive knowledge, and suggest that its ever-inquiring character, its restless skepticism, is its most basic attribute. It is possible to present the breakdown as not the product of science but of its political misapplication. And one could observe, philosophically, that given a couple of decades the scientific community had shown itself capable of righting its errors. The misreading of fire was simply a longer example of fads like cold fusion and polywater, which the community ultimately self-corrected. In the long run, science was right. Yet a land agency is not a research institution: it must act, it needs workable knowledge to perform its tasks, and the consequences of error cannot be overturned by the latest journal article. Besides, in the long run, as John Maynard Keynes famously observed, we are all dead.

The reality would seem to argue that local knowledge based on centuries, if not millennia, of practical experience coded into cultural mores was far superior to field- and lab-generated (and later, computer-simulated) data. It just didn't have the same cachet, and it could threaten to undercut the claims to privileged knowledge that led to money and power. Interestingly, the Clark's Peak, Gibson, and Nuttall fires did not obey the fundamental logic of fire behavior by which the most vigorous burning would occur upslope. The steepness of the massif apparently encourages very strong temperature gradients as, at the end of a day, the top cools rapidly while the shoulders remain heated and a violent sundowner wind results. The worst blowups were actually blowdowns. Abstraction met local circumstance, and the facts won.

The general response has been to do more of the same. The solution to the problems of science is more science of the same sort. That the agencies responsible for administering the land are also the ones sponsoring research means that there is no way to segregate the two. Science cannot exist apart from politics because politics pays for the science. It cannot merely observe and analyze from a neutral vantage point because those operations and the vantage point itself deform the scene being examined.

The reality, too, is that major reformations in fire management have come not from new scientific discoveries but from changes in cultural values. Upheavals in social understanding determined the paradigm shifts in fire science, not vice versa. Critical thinkers came to value fire because they saw it as part of wilderness, not because they chronicled its evidence in scarred trees and soil charcoal. Those cultural revolutions further allowed society to sift through the competing claims of the various sciences. The ideas and beliefs that surfaced chose which kind of research to support and which to put on the shelf. Still, perhaps damningly, the most influential publication of recent decades, Norman Maclean's *Young Men and Fire*, was not written by a fire scientist but by a professor of

Renaissance literature at the University of Chicago. After the disastrous 1994 season, the book directly affected the adoption of a common federal fire policy and helped convince fire officers to fight fire differently, one consequence of which has been a willingness to trade burned acres for enhanced firefighter safety.

After nearly a century of evidence, it should be clear that fire science is not adequate to the task of full-spectrum fire management, and that it will never be adequate. Science, as science, simply can't answer the questions most needed to live on the land, which lie in the realm of cultural values, a moral universe impermeable to the lens of modern science. It can improve technology and advise about possible outcomes of decisions, it can overgrow with data, but it cannot decide, and its record is such that acting solely on its existing data will almost certainly lead to errors if not disasters. It should be one scholarship among many, and one epistemology among a throng that includes the impossible-to-codify-and-reduce-to-numbers experiential reality by which people actually live. Yet there it resides like the MGIO, demanding ever more space to do what it deems essential, insisting that it sits above criticism, willfully agnostic about the scientific-industrial complex that supports it.

The critics of fire suppression often point to graphs of increasing expenditures and swelling acres burned to make a case that more money fighting fire doesn't reduce either costs or burned area. Defenders will reply that worsening conditions—climate change, the WUI—are determining the fundamentals, and that these deep-driving circumstances are causing the megaburns that bring larger suppression costs. Yet, in the perverse way of correlations, critics could impishly hint that the rising expenditures are just as likely to be the cause of increased burning. The more we spend, the less control we get. A fire suppression–industrial complex is pushing up costs without regard to results on the ground.

This same logic can be applied to fire science. An uptick in fire research parallels the same upswings in firefighting costs and burned area (graph on page 93). New sponsors, new fields, new shelves of science. The USGS has joined the USFS as a funding agency, and the Joint Fire Science Program, established in 1998, has pumped significant monies into research. The number of scientific articles published shows an exponential rise: in the early 1960s some 13 papers per year were published, and in the early 2000s over 300. Partisans will argue that the growing crises, worsening

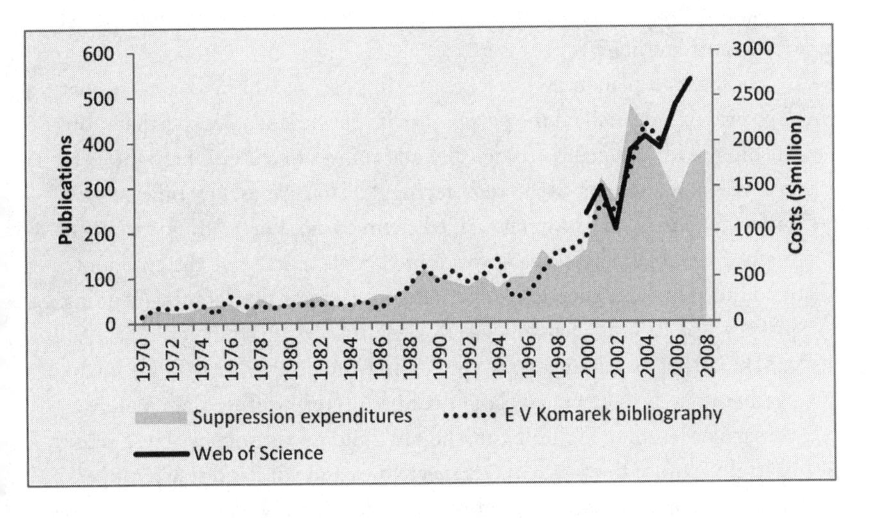

circumstances, and emerging megafires are the reason for more research funding, and that the proper solution is still more funding for still more studies. Yet this is exactly the logic that long governed suppression. One could just as easily argue that the enhanced investment in science has not made any difference on the ground, or even that an emphasis on fire sciences has diverted attention from the real "drivers" of fire's management. An objective measure of applied fire science—analyzing science as science would natural phenomena—would probably show mixed results much like that from fire suppression. The more we spend, the fewer practical outcomes we get. A fire research–industrial complex is pushing up costs without regard to results on the ground.[3]

Apologists brush off such observations as they might an annoying deer fly. They know that authority goes to power, power goes to money, and the money goes to science. They might further demur that science only observes and analyzes with complete disinterest. In fact, science can deflect from other forms of inquiry and by counseling practitioners it actively alters the landscape it studies. Over the years it has measured a landscape shaped by decisions informed by past science. It has affected the Pinaleños as fully as the Mount Graham International Observatory. This places science, and its institutions, squarely on the summit; and like the squirrels it has nowhere else to go. If it affects the outcome, then it is part of the problem. If it has no effect, then why is it granted special status and funding?

A better explanation for increases in cost and burned area is that America has reclassified the purposes of its public lands, accepts that more burning is advantageous ecologically, and refuses to commit firefighters to go mano a mano with fire in remote, rugged landscapes. Fire officers back off, as federal policy has encouraged them to do. For 30 to 40 years the major federal land agencies have adopted goals to increase the amount of land burned under their care. The statistics suggest they are finally doing just that. Be careful what you wish for—and how you study it.

The stronger argument for supporting fire science is that fire management needs to engage its larger culture on terms other than merely as vernacular learning and folklore; the fire guild needs, somewhere, a sense of itself as more than backwoods mechanics and wildland sharecroppers. It needs to connect to high culture in order to truly engage its sustaining society. It needs sophisticated fire science for the same reason that a modern culture needs astronomical observatories. The difference is that observatories tell us little about how to manage Earth, while wildland fire science intends, as its announced ambition, to influence conduct, which is to say, to shape fire practices on the ground. But neither is without cost. Those observatories compete with other values—can literally shove them aside; they claim, and defiantly occupy, the high ground. So, too, fire science can push to the margins other scholarships and forms of knowledge.

Those scopes on the summits will not melt away: their internal imperative is to expand. To even appear to criticize science is to invite charges of philistinism, politicization, and capitulation to faith-based superstition. A steely-eyed survey of fire science's actual achievements, however, would point to marvelous insights into nature and a much-flawed record of practical outcomes. And that, in the end, makes the wildland-science interface a far more troubling conundrum for the fire community than the better publicized wildland-urban interface. It's easier to defend those trashy trophy homes than to dismantle telescopes.

Yet, unexpectedly, the imperial model of science, in which science informs and management applies, is finding itself constrained. Nationally countermoves are underway in the guise of traditional ecological knowledge and of adaptive management that blur the hard border between science and quotidian experience, in which science becomes experiment in management and practice becomes a scientific experiment, with both needing to be constantly calibrated, compared, and adjusted. Granted some space the concept may return fire management to its ancient status as grounded in experiential knowledge. In principle it restricts science to standing as one form of information and political input, much as modern fire management identifies suppression as one option among many.

In the Pinaleños the general must always interact with the specific; and here, the alchemy by which principles meet particularities can yield unexpected outcomes. In this case the specifics are the reality of the Mount Graham International Observatory. While the University of Arizona and big science got their way—got exclusive rights to the site, built an edifice like a Borg cube that can be seen in reflected sunlight from the White Mountains to the Mexican border—they find themselves denied any larger claim. The road and facilities occupy 8.16 acres. That zone is fenced by stakes and a yellow acrylic rope, beyond which MGIO residents are not permitted to go. The grounds are patrolled by security officers and dogs; they are there to keep in as well as to keep out.

Ask anyone about Mount Graham, and you will be told the place is "political." Of course it's political: it should be. These are public lands and arenas for public values, and in a democracy politics is where competing values must be openly discussed and decided. What tainted the MGIO was that its politics was not open and honest. And what has compromised so much of fire science in the past is that it has confused its data with its values and has dismissed any other scholarship and any other competing values. Science there must be. But it has no claim to the whole of the summit.

THE VIEW FROM TANQUE VERDE

STAND ON TANQUE VERDE RIDGE in Saguaro National Park and see, in one compelling panorama, all that makes fire management in the western United States problematic. A metropolis lapping at its gates. A Class I airshed tainted by industrial pollution and reluctant to add more. Invasive grasses. An isolated mountain range layered with biotas and stirred into varied fire regimes. Endangered species. Exposure to public scrutiny. A service economy. An immigrant population drawn from fire-immune places and a resident population spiked with academics. Habitat fragmentation. A forest piled with fuels like a landfill. Prodigal lightning, assuring that a landscape primed for fire will always have ignition. They are all here, and they converge with the majestic syncretism that characterizes the Madrean archipelago.[1]

On closer inspection the scene worsens, if that is possible. The city is Tucson, one of the Southwest's Big Four, spreading like adobe kudzu through desert, wash, and foothill. Between them the Rincon Mountains and the Santa Catalinas, meeting to the northeast, form two sides to the Tucson basin, making a gaunt amphitheater. Invasives began with red brome, a flashy pyrophyte, in the 1970s, but the cycle of drought beat it back, and instead buffel grass, a pyrophytic perennial, much denser with calories and far more tenacious, replaced it. In the mountain forests the Mexican spotted owl complicates efforts to prescribe burn or conduct wholesale fuel treatments, while in the Sonoran Desert fire can undermine the habitat of lesser long-nosed bats and leopard frogs through burning and postfire flooding, and flames directly threaten the park's signature species, the saguaro cactus. Settlement from mining, homesteading, and ranching had wrecked the indigenous fire regimes in the usual ways by constructing roads, trampling, stripping off the grasses, and sweeping aside the native burners; grasses morphed into brush and woody litter. In 1902 the mountains were gazetted into the Santa Catalina Forest Reserve, subsequently reorganized into the Coronado National Forest, which imposed a program of active fire suppression that Saguaro National Monument inherited in 1933. The monument itself is split into two units, an eastern one atop the Rincon Mountains and a western one on the other side of the metro area amid the Tucson Mountains (proclaimed in 1961), such that habitat fragmentation extends even to the political composition of the park itself. By midcentury a rural economy that had used fire where it could amid the pummeled grasslands was gone, replaced by a service economy of tourism, government programs, shopping malls, military bases, a university, and retirement communities, none of them eager for smoke and flame.[2]

Any one of these factors could have resulted in a cautious or compromised commitment to restoring fire, or in its effective extinction—and has in plenty of places. A major failure could have cut the program off at its knees. A failed fire could be quietly buried in the obscurity of the Mimbres or Mount Trumbull. In Tucson the mountains would act like a megaphone. Too many things could go wrong. As the folk saying has it, it's better to be lucky than good. A fire program at Saguaro would have to be both.

Yet, paradoxically, those factors that seemingly worked against a fire program also made it work. The small can be more nimble than the large, and an administratively remote diamond in the rough can have more freedom of movement than the closely watched crown jewel. What in other times and places might be liabilities could become assets.

The park's partition, for example, segregated its two fire problems. The western, Tucson Mountains unit, was desert and saguaro, and sought fire exclusion. The eastern, Rincon Mountains unit, pursued fire restoration. The mountains had Sonoran Desert and saguaro-clad foothills, some infected with invasives, and fires in these lower elevations it actively suppressed. Fire on the summits had a harder time burning down crenulated slopes, and prevailing southwest winds drove flame (and smoke) away from the saguaro sanctuaries and Tucson. Much of the region's tourist and snowbird population filled the basin in the winter, while its fires followed the summer monsoon. Its population was educable. The University of Arizona hosted the Laboratory of Tree-Ring Research, a world-class facility, and boasted a strong natural resources faculty, also interested in fire. During the 1970s, when the park began experimenting with new fire practices, Tucson evolved into a literary center, many of whose writers (like Ed Abbey) were attracted to environmental topics. The public granted the National Park Service more freedom of movement than it did Saguaro's surrounding Forest Service districts. And the NPS allowed its managers a degree of autonomy unknown among other federal agencies; the right ranger could move quickly to install a program. The monument (boosted into park status in 1994) never experienced a blow-out fire that attracted the angry attention of national authorities or drew the wrath of its watchful publics.

Fire exclusion had obeyed the classic Southwest cycle, gaining purchase in the late 19th century. A serious if minimal fire-suppression campaign by the Forest Service had started in 1922. Horseback patrols continued through 1940, when the monument assumed responsibilities (official records begin in 1937). Control was weak during the war years, and relatively big fires still burned into the mid-1950s. Saguaro introduced its program for restoration in 1971. Its first prescribed fire came in 1984, followed by nine more. The age of active suppression by the park had lasted perhaps 30 years; by 2011 the age of restoration had run 40. As a protected site Saguaro had experienced more time under the new regime than the old.

The park never achieved what it intended—no place ever has. But it had held at bay the pressures to keep fire out, it had a functioning new-order program, and in the early 1970s it briefly shared the national stage by operationalizing the concept that would, after endless iterations and linguistic metamorphoses, become the foundation doctrine for fire's management in western wildlands, what its promulgator, Chief Ranger Les Gunzel, called a "natural prescribed fire." It was the second unit in the national park system to do so.[3]

The immediate inspiration was the Leopold Report that the NPS encoded into a new handbook of administrative guidelines promulgated in 1968. Sequoia-Kings Canyon jumped on the prospects early and had a "let-burn" in its backcountry that summer. Most parks (and a few forests rich in roadless areas) pondered how to transition to the new regime. America's great cultural revolution in fire watched a hundred flowers bloom, as smokes drifted upward from the Selway, the Gila, and the Sierra Nevada.

In 1971 Saguaro inaugurated a program to restore lightning-kindled fire with the announced goal of "producing a fully natural area for future generations." Lightning started 11 fires that summer, of which the park let 10 run "their natural course." By 1974, when the park published its first formal fire plan, it had permitted 24 of 46 fires to burn freely. Overnight, it had, gestalt-like, moved from suppression to restoration, as though a picture long viewed as a vase now appeared to resemble two faces. All it took, it seemed, was a switch in perception. The new order didn't require the park to do anything new, only to stop doing what it shouldn't have done in the first place.[4]

The Saguaro solution synchronized philosophy and practice. The philosophy was easy. The charge to the park was to preserve and protect the natural scene. The fires would come and go with the rhythms akin to those by which saguaros blossomed or deer moved up and down the mountain with the seasons. Sense and policy argued that nature should rule, that the wisest fire policy would be to back off, as park officers did with other natural processes. If a fire went rogue, like the occasional bear or cougar, then it would be hunted down. But the presumption was that it belonged. The park's true fire problem was not wild fire, but fire's suppression. There was no more justification for beating down those flames than for continuing to kill coyotes. The arguments strengthened in 1976 as 90 percent of the park went into formal wilderness.

Saguaro's physical geography favored the same solution. The Rincons were roadless, and trails were few, which made access onerous. Natural barriers were sparse—little more than what the abrupt, crinkled slopes of the massif offered. Prescribed fire could not rely on internal buffers or a resident population. Every day might be a burn day, but if no one was on the mountain, no one could ignite the burns, and without creating an expensive infrastructure for containment, those ignitions were no more under control than if lightning had set them. The simplest solution was to massage natural conditions into a prescription and let nature do the burning.

The natural fire plan claimed only a handful of pages. Its operational guidelines were few and expressed in plain-text language. Suppress human-caused fires, fires around structures, fires along the border, and all fires within saguaro patches. Eliminate mechanized firefighting except where required to save life and buildings. For the monsoon season, from mid-July to mid-September, once two inches of rain had fallen at Manning Camp by Mica Mountain, fire would be left alone. There were some limits imposed by considerations of predicted fire behavior, which could push fires ignited under extreme conditions into suppression. But the average fire in the average year would no longer be hunted, trapped, or poisoned.

Within a handful of years a remote monument of 91,327 acres, of which 47,000 were fire adapted, went from obscurity to national prominence. Half of Saguaro's lightning fires free-ranged. They burned as predicted; along the ground, sweeping away the litter, unthreatening to sky and city. The "natural prescribed fire" (later, transposed into "prescribed natural fire") merged the two protest movements that had successfully stalled the suppression juggernaut. It was prescribed—that is, desired, scientifically grounded, and controlled; and it was natural, as a wildland fire should be. Appropriately, the fire plan was published as an appendix to a general landscape management plan.

Surely, it was too much to expect that this momentum could continue. The park's mission was its saguaros, not its ponderosa pine and Emory oak. The liberation of national park fire programs, flourishing like fireweed in the aftermath of the Leopold Report, would inevitably be succeeded by the bureaucratic equivalent of more shade-tolerant species as the NPS sought both to free and to contain its fire experiments—a paradox not unlike the prescribed natural fire. Once the euphoria passed, the reins tightened.

Further addendums were added in 1974 and 1978. A formal fire plan was published in 1979, along with the return of a national manual (NPS-18), followed by an interim update in 1983, and in 1992 a full-spectrum revision written to evolving national specifications and that absorbed post-Yellowstone lessons. By then the program, despite greater investments, was ebbing. The trend was evident even as the park crafted that first stand-alone fire plan in 1979. The document tabulated the record to date: 40 natural fires and 918.5 acres burned. Twenty of those fires and 896 acres, however, had occurred in the first two years. Subsequent seasons never reached anything like those proportions. Fewer fires were allowed; with a couple of exceptions they burned fewer acres; and prescribed burning was unable to make up the difference. The park seemed to recruit fires as the Sonora Desert did saguaros; its fire program waxed and waned as conditions warranted.

The program stuttered in the field for all the familiar reasons: what had pushed it ahead now pulled it back. The fire organization had persisted, and with it the tendency for suppression to become a default option amid ambiguity or when conditions were less than ideal. The prescribed natural fire program stalled after the 1988 Yellowstone fires caused a systemic reboot, and the same held for prescribed fire after the 2000 Cerro Grande debacle. Prescribed fire proved too complicated and expensive to substitute for lost natural fires. The experiment reached a breathtaking climax in 2010 when the 110-acre Mica Mountain burn cost $300,000; such monies, allocated under the National Fire Plan fuels-treatment projects, would no longer be allowed away from the wildland-urban interface. The 4,500-foot elevation that roughly separated the desert from the woodlands—and that segregated banned burns from tolerated ones—was arbitrary, and difficult to enforce. The troubling legacy of stockpiled fuels argued for a prudence that could easily segue into pusillanimity. Buffle grass promised to reconnect what natural and recent history had fragmented—a shotgun marriage that would allow fire to spread across what had formerly been natural barriers. Besides, a single wildfire could consume the entire park. The 2011 Horseshoe II in the Chiricahuas fire burned four times the fire-adapted area of Saguaro; the Wallow fire in the White Mountains, 12 times. One feral burn could break the bank.

Then there were the actual fires themselves. The earliest had behaved exactly as advocates hoped. But when a natural fire in 1994 threatened to cross the park boundary into national forest, the Forest Service discovered it had no funds for "prescribed fire" and no means to jointly manage such a burn, and the fire was suppressed. The 1999 Box Canyon fire, powered by red brome in the lower foothills, burned into saguaro stands, effectively doing what urban developers had been prohibited from doing. The Helen II fire in 2003 crowned and wiped out prime Mexican spotted owl habitat. When the desired landscape fires had come, they had not behaved as expected. While subsequent landscape-scale planning among the agencies has eliminated some of the administrative concerns, it has not restored the expunged fires. Reforms take time, and time translates into missing acres burned.

Everything has become more cumbersome. What in early plans had been typewritten, or hand-annotated appendices, is increasingly encased in an exoskeleton of bureaucratic procedures. The 1971 appendix totaled 12 pages; the 2007 plan, stuffed with boilerplate legalese, ran to 157. As planning to accommodate fire swelled, actual burning on the ground shriveled. Local lore, local control, local autonomy—all yielded as the complexity of fire on the American scene argued for greater political considerations, which meant more encumbrances, which moved liability and power upward. Knowledge of fire behavior and effects gained by long-tenure ("homesteading") rangers was replaced by GIS data banks, commissioned research, and LANDFIRE. The pioneers passed from the scene. The 2007 plan was written by a professional contractor.

For a while it seemed that climate and notoriety might rekindle the program. The Southwest swung into a cycle of drought and lightning starts, while after the 1988 Yellowstone fires, the NPS found itself flush with fire monies, to which the National Fire Plan added more. Instead of a hotshot crew, Saguaro housed a fire use module, ready to attend to prescribed natural fires. A 2004 summary observed brightly that over the past decade some 12,000 acres had burned as wildfire, 10,000 had free-burned as prescribed natural fires, and 3,600 had been prescribe burned. That year's budget exceeded $850,000, apart from emergency expenditures on wildfires. It was a lot of money per acre burned. The founding premise had been that free-ranging fires would be cheaper and safer than fought ones. Although advocates could point to success in principle and nearly half the park had burned in some form during the previous decade, skeptics could note the inability to keep up with original promise.

To them the program appeared to be afflicted with a kind of wasting disease, withering neuron by neuron, limb by limb.[5]

The Rincons had in presettlement times burned every 10 years or so on average. One cycle, two cycles, maybe three cycles might be skipped without enormous distress. But the Rincons had missed roughly five return intervals prior to establishment of the monument, and some eight since then. The new era averaged 100 to 200 burned acres a year. With 47,000 fire-adapted acres the park was morphing from a 10-year fire cycle to a 313-year one. Restoration had, at best, introduced a single burst of burning that did not even cover the landscape once fully. Besides, what mattered ecologically was not the return of a burnover, but the reestablishment of a fire regime. It wasn't happening. In bulk terms, fire continued to leach away from the mountains. Combustibles still ratcheted upward like the speculative frenzy that precedes a stock market crash. To the minds of at least some observers nature's economy was poised on a cliff.

It may be that the issue is less about bureaucratic barriers than the problematics of size and the probabilities of timing. It may happen that recent endeavors like the 2005 interagency merger of Santa Catalina and Rincon fire management and the FireScape project will allow for the necessary economies of operational scale, permitting Saguaro to share expenses and liabilities with the Coronado and the regional fire community. More critically, the clarification of what "appropriate management response" or its latest iteration and avatar ("strategic management response") means within federal fire policy may replace the troubled prescribed fire programs with a more flexible and readily funded response to wildfire, a variety of boxing and burning. Fire will be fire, as the mantra goes. Burned acres, partisans hope, will follow. Managed wildfire will get the acres prescribed fire could not. Maybe.

Or maybe not. The park's signature species renews itself episodically. It recruits when conditions are favorable and lies dormant when they are not. Each flower blooms for 24 hours, with the bulk of its pollination done at night by bats in what seems a miracle of ecological synchronicity. A new generation must then establish itself in the ground. A saguaro cactus lives 150 to 200 years, which is to say, the old-growth saguaros that justified the park were established in the 50 years prior to the Mexican War and Gadsden Purchase. The giants that flourish today germinated about the time of Anglo settlement; they declined as cold snaps and fires and the various insults of settlement, particularly grazing, visited them; they have rebounded when those pressures have lifted, as they have recently. Yet while long lived, the saguaro is notoriously shallow rooted. For all its iconic majesty it is an opportunist.

That improbable matching of pollen and pollinator, and of seedling and survival, might well stand for what has happened with Saguaro's fire program. It flowered briefly, blossoming during a brief pollination when people and climate converged. The program took root, then pulsed through dormancy and revival, at times robust and often faltering as conditions opposed further recruitment. Now it must find ways to persist and grow a new generation in the face of an increasingly inimical setting. Fire and manager will have to meet within whatever brief efflorescences of opportunity occur. Saguaro will have to be lucky as well as good.

FIGURE 1. The Huachucas: the view southeast from Montezuma Pass. Burnout operations protected most of the ravine, which held the memorial's visitor center and had biological value as a riparian zone. The border fence is visible in the background, upper center.

FIGURE 2. The wildland-science interface at Mount Graham. The road leads to the telescope compound. The rope defines the border between scopes and squirrels.

FIGURE 3. Wally Covington looking at the Warm fire. Here it burned through mixed conifer and is regenerating into aspen. Wally's concern lay more with old-growth ponderosa pine for which this kind of conflagration was well outside evolutionary experience.

FIGURE 4. Bonita Creek Subdivision, 2014. To the left of the photo the 1990 Dude fire overran and killed six firefighters. Houses were rebuilt, and a cordon of clearing extended into the scrubby regrowth; a second pass, in foreground, is underway. The Mogollon (Tonto) Rim is visible in background.

FIGURE 5. Cochití Canyon, Jemez Mountains, three years after the Las Conchas fire swept through like a torrent. What had been a dense mixed-conifer corridor held only tiny patches of green.

FIGURE 6. Aftermath of the Wallow fire as it curl-burned over the rim around Alpine. A wind-driven fire, the burn scar ran around the valley like a bathtub ring.

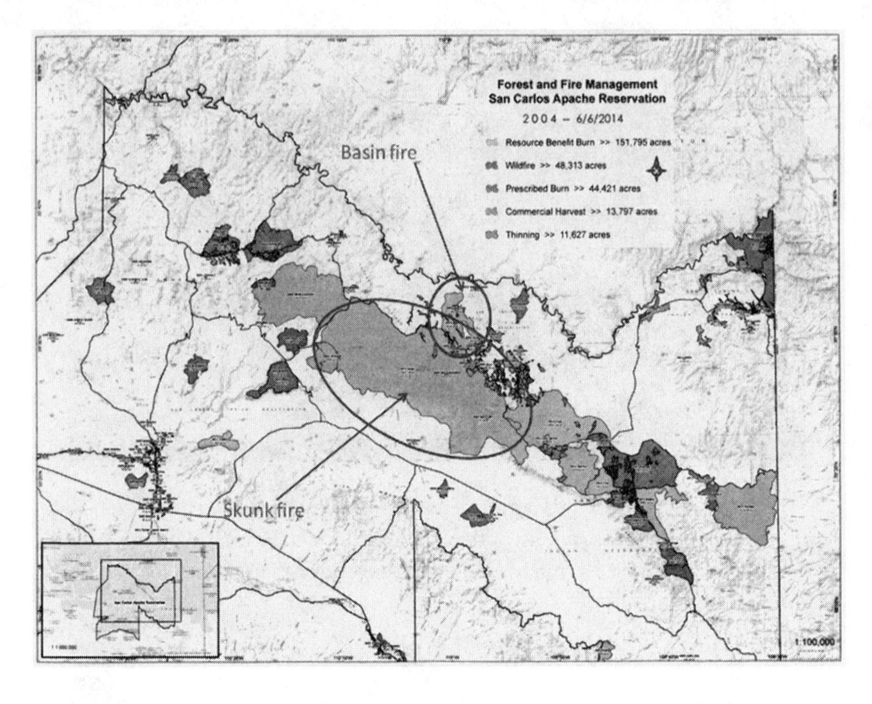

FIGURE 7. Fire atlas for the San Carlos Agency, showing the 2014 Basin and Skunk fires. Courtesy San Carlos Agency; modified to indicate recent burns.

BORDERS AND RIMS

WITH ITS SHARP edges and sudden shifts—rims that plunge into gorges, mountains that rise from plains, mesas that etch against the skyline—the Southwest's terrain bristles with borders. There is little subtlety about their edges. The landscape fragments into chunks and tiles that, like a mosaic, make a coherent picture when viewed from afar but shatter into pixels when viewed close up. Because those patches define fuel arrays and fire behavior blocks, they influence how fire flows. It's a question of scale. Small fires spread within small blocks. Big fires overrun the small edges and swell to the proportions of big blocks.

This fact suggests natural units for fire management. But superimposed over those tiles are the cycled layers of human settlement. Tribal homelands. Private ranches. Cities. Exurbs. National parks, wildlife refuges, national forests, national conservation lands, state trust lands, national monuments, bombing ranges, military proving grounds, Nature Conservancy holdings, all overseen by distinct institutions and all nestled within or sprawled across political jurisdictions from municipalities, counties, states, and the federal government. The Southwest is awash with borders—some hard, some soft, some impermeable, some porous. A few of them fire must acknowledge. Many it will ignore. But they are all too real for fire's management.

REINVENTING A FIRE COMMONS

YOU HEAR THE SAYINGS EVERYWHERE. Whatever the encumbered language of fire policy might proclaim, practice increasingly pivots around two mantras: fire is fire, and fire does not respect borders. Together they epitomize fire modernity. They encapsulate a half century of the fire revolution, as they seek to unburden fire management from the phony distinctions of a fire discourse that has seemingly descended into a scholasticism of arcane terms. They allow fire room to roam. They give fire officers space to maneuver. They permit fire management to operate on a landscape scale.[1]

Yet both mantras are half-truths. Fire ignores boundaries that are jurisdictional lines on a map, but it certainly respects any border that changes its behavior. There are many fires above the Korean DMZ and few below. The Great Plains display a checkered pyrogeography because of roads and plowed fields. A shopping mall surrounded by asphalt will stop any fire from the adjacent woods. The whole premise of the WUI is that borders matter. Even more, the mantras are not as original as believed, for they recapitulate in modern parlance the world that existed before the revolution, the outcome of a long evolution from the Weeks Act in which cooperative programs and mutual aid agreements created a fire management commons under the benevolent despotism of the Forest Service. Then, too, fire was fire because all fires were destined for suppression, and fire did not respect borders because suppression required the ability to move crews and equipment to where the need was greatest. Decades of protracted effort glued those fractious shards under a common purpose. They allowed fire protection to operate on a landscape scale.

What matters today is that the earlier world broke apart, and it has taken nearly four decades to put the pieces back together. The fire revolution was not about restoring fire per se: it was about aligning fire management with land management. Since different lands had different purposes, they required different policies, different practices, even different research apparatus to meet their charter. National forests and grasslands and BLM lands, all of which had multiple uses, found that the public wanted different uses for them; and where wilderness was at stake, those uses were incompatible. All these classes of land brought with them new categories of fire, each with prescriptions and proscriptions for use. One fire regimen could not satisfy them all. The revolution, that is, required that the fire commons be enclosed and parceled up as a precondition to a modern economy of fire management. It demanded borders.

Over time, those borders, both jurisdictional and conceptual, have become an encumbrance. They don't stop fire; they do inhibit fire management. They aren't the landscape in miniature, just landscape in pieces. The fragments are smaller than the fires. Once again, the fire community finds itself searching for ways to blur or erase those borders and install a new commons.

Reassembly began with suppression; with concepts like total mobility and with institutions like the National Interagency Fire Center and the National Wildfire Coordinating Group. It spread to prescribed fire with the creation of common protocols and certification. It embraced research with the Joint Fire Science Program. It bonded with civil society through the Nature Conservancy's Fire Learning Network. But fire restoration, in particular, remains event driven. Prescribed fires require site-specific plans and times.

Strategists realized that the only realistic way to reinstate fire on the scale required was to exploit wildfires, to redirect them to more benevolent purposes. Here, however, the origins of the fire revolution in differing land use, which differentiated among fires and responses to them, became an impediment. The national firescape was a bestiary of wild, prescribed, planned, unplanned, fire-use, confined, contained, and other fires; and effective management meant that these conceptual divisions had to undergo the same reorganization as the land so that there were fewer kinds of fires on a larger management setting.

The only solution, as nearly all parties acceded, was to operate on the scale of landscapes. Fire's reinstatement needs the same scale as fire's suppression and that scaling must encompass organizations as well as mountains. Landscape fire management, that is, requires the ability to organize the institutional landscape, to build in buffers and decision spaces, to prevent polarization and stalemate by allowing many competing interests, to treat suites of projects under a common process for political sanction and NEPA review.

What makes a shared institutional landscape possible is not only a common fire policy but a commonality of purpose—almost all federal lands now seek to reinstate fire, or to check wild fire by substituting less savage versions. But what moves landscape beyond the public agencies is the sense of impending crisis from large, severe fires that serve no one's interests. Fire isn't waiting until all the studies are in, the competing parties negotiate away their differences, and the public wearies of celebrity gossip and attends to matters of civic substance. They are burning more fiercely across ever-larger landscapes. Fire management has to respond in kind. It needs ways to light fires, watch fires, and herd fires on the same scale it has traditionally fought them.

Even as the fire revolution evolved, arguing for a pluralistic perspective on fire, America's national estate was polarizing into the wild and the urban. The wild (and free) was the antithesis of the built (or suppressed). For some 20 years American fire management struggled over how to cope with fire in the growing National Wilderness System. For the next 20 years it has grappled with fire along the urban fringe. What was lost in the upheaval was the vast middle landscape between them, which was also potentially a middle ground. That's where the Four Forests Restoration Initiative (4FRI) resides.

A long introduction—but then the Four Forests Restoration Initiative has had a long gestation. Its geographic profile places it along Arizona's

Mogollon Rim, amid four national forests that hold the largest belt of contiguous ponderosa pine forest in the world. Its charge is to promote "a collaborative, landscape-scale initiative designed to restore fire-adapted ecosystems" across 2.4 million acres of ponderosa pine on the National Forest System. It has become the national pilot project for NEPA, and the Forest Service's top priority under the Collaborative Forest Landscape Restoration Program. Critics fret over whether it is too big to fail or too big to succeed. But scale is its essence. The belief grows that individual projects can't keep pace with the problem. They don't add up to a whole.

Why here? Because geography and history have converged amid a sense of crisis. The montane forests, particularly of ponderosa, are adapted to routine surface fire and have been especially unhinged by fire exclusion. This history is well understood. Twenty years earlier Wally Covington had begun research on treatments that could restore the forest to something like its presettlement savanna state; those trials evolved into a demo plot and a method of social engagement that became known as the Flagstaff model, which then underwrote thinking on the Healthy Forest Restoration Act. The fire community understands how, in principle, to reinstate fire in ponderosa; and once the grassy understory is restored, maintenance is relatively simple.[2]

Meanwhile, appreciation swelled among all thoughtful observers that the present landscape was a mess, fit only for wildfire, which began arriving in syncopation. The Mogollon Rim experienced spasms of horrific crown fires, well outside the evolutionary experience of its pines and beyond the capabilities of fire officers to keep from invading exurban communities. In 1996 the Horseshoe and Hochderffer fires blasted the mountains around Flagstaff. In 2002 the Rodeo-Chediski burn metastasized into Arizona's largest forest fire of record. It led to a governor's Forest Health Advisory Council and Forest Health Oversight Council, which sparked a 20-year strategy to "restore forest health, protect communities from fire, and encourage appropriate, forest-based economic activity." The outcome was the 2007 Statewide Strategy for Restoring Arizona's Forests. When Congress passed the Collaborative Forest Landscape Restoration Program in 2009, the state's fire community had the science, the political will, and the sense of urgency to propose a very large project that would extend over the Coconino, Tonto, south Kaibab, and Apache-Sitgreaves National Forests. The Four Forests Restoration Initiative was the result.[3]

The basic strategy is to fashion conditions to reinstate fire in something like its evolutionary role. That means treating swathes of 5- to 18-inch dbh ponderosa by slashing, crunching, or otherwise removing it on strategic sites. The plan calls for treating over 30,000 acres a year of such "over-represented" woods. It also means letting fire further its own return on 60,000 acres a year. Natural, prescribed, wild—the categories of fire matter less than the space granted to them. Mechanically treated sites, old burns, and new prescribed burns will fashion a firescape in which active fires can be seen not as a threat to be attacked but a process to be encouraged and herded.[4]

Critically, the scheme also expands the institutional landscape. Some 31 stakeholders in addition to the Forest Service have signed on. They range from the Northern Arizona Wood Products Association to the Sierra Club to Gila County to the Arizona Wildlife Federation. The fundamental problems are not technical issues of silviculture and ignition patterns, which can be resolved by scientific experiment. They are cultural and social, how to reconcile so many competing interests on a given patch of public domain, which can only be managed, not resolved, through political experiment. 4FRI is a deep exercise in participatory democracy at a time when polarization and stalemate have become the operative terms for national politics.

The Forest Service remains the responsible agency: these are public lands that it manages on behalf of the nation. But while it conducts the planning and oversees the work on the ground, stakeholder committees do the heavy lifting regarding acceptable standards and guidelines, each from their own perspectives, yet in such a way that they do not override other publics or subvert Forest Service authority. Without dissolving their own identities, participants join a larger cultural landscape, of which each represents one tile. Nor does 4FRI void the NEPA process. Rather, it seeks to expand it to match the landscape scaling of the work and institutional mosaic. As it seeks to move beyond fire suppression, so it aspires to move politics beyond the courtroom. If it fails, then wildfire and litigation will return. The agency will be fighting fire in the field and the community fighting one another in court.

Wildfire doesn't care; it will happen with or without agreement. If it is not simply to react, 4FRI must match fire's inevitability with its own. It had to move beyond studies, planning exercises, and talk-shop meetings and act. It planned to go operational in 2012. It missed that target through fumbled contracts. In November 2014 it released its Draft Record of Decision for the Kaibab and Coconino forests initiatives. The decision applied to 586,110 acres and established a model for proactive fire management. The remaining question was whether 4FRI could evolve faster than the four forests' fires.[5]

4FRI is a work in progress. Part of its founding genius, however, is to recognize that it will always be a work in progress.

The Four Forests Restoration Initiative has no goal other than to prepare fire-adapted ecosystems to receive fire on something like their historic scale. It accepts that no person or entity holds the singular truth about what that ambition means. It pursues the "best-available" science with the understanding that science is ever incomplete (and although it does not say so, with the implicit recognition that the errors of the past have been promulgated with the imprimatur of the best-available science of their times). What, ultimately, it restores may not be a particular landscape but a style of American pragmatism in which it is possible to act amid a contingent world about which social groups have incomplete and competing knowledge and claims over what it means.

That, in fact, may be the political pivot. The program is charged to *do*. Stakeholders can sign on to help, or they can stand aside. Compromise is inevitable, but compromise can come without consensus. 4FRI does not lead to conversion: it requires no agreement on first principles or creeds, only on what is to be done. Each confessional group can refract that deed through its own interpretive prism. The sheer multiplicity of participants prevents the kind of polarization that often paralyzes. It helps that legal wilderness is off the map and that that cities are eager for protection through treatments. The wild and the urban don't face each other across an impermeable institutional interface.

It may be critical that only one landowner is involved, the Forest Service, which has responsibility for actual implementation; that the landscape, while minced by jurisdictions, is ecologically simple and uniform; and that all parties agree that fire must return. Interestingly, the experiment recycles the Forest Service to its historic role as the indispensable agency for fire. It alone has the whole apparatus needed for management, even if it is downsized or its parts are thinned into ghostly vestiges. It is the one institution that all the others connect to—the one agency that integrates the institutional landscape as ponderosa pine does the ecological.

In establishing its enlarged scale for fire management, 4FRI softens interior borders. Fire management, not just fire suppression, can cross boundaries along with fire. The blurring of institutional lines relies on a useful ambiguity, one historically common to American politics and the philosophy of pragmatism: the substitution of process for product. American jurisprudence doesn't begin with a definition of justice, which a pluralistic society not bound by blood and soil could never agree on: it defers to due process. So fire management doesn't demand agreement on a definition of what the land should be. It is enough that treated places will adapt as they move toward greater resilience in the acceptance of fire. Of course agencies must specify targets for budgets, critics demand anticipated outcomes for accountability, and NEPA approval requires identifying a "desired condition."

Once the process names a particular action, it hardens, and the trail becomes littered with stones. Each stone may seem large to an individual stakeholder, so much so it may block the trail. They seem small only if one looks at the landscape overall. But that is the perspective that 4FRI demands. To restate the issue in terms of pragmatism, a decision doesn't demand consensus over the ultimate result or even over its interpretation, only over the action taken. The truth of those separate ideas doesn't matter, only agreement on what to do next. As William James, one of the architects of pragmatism as a formal philosophy, put it, By their fruits ye shall know them, not by their roots. Something is true or not according to how it survives real-world tests.

What emerges is a reconstituted working landscape—not one, as before, committed to resource commodities, but to the production of ecological goods and services, and a place where people can act. It is a new avatar of the old multiple-use vision in which aesthetics, biodiversity, ecological integrity, and recreation replace fiber, forage, water, and ore. In both visions fire remains integral. This time, however, the intention is to enhance ecological communities through good fires rather than to protect commodities from bad ones. The ecological working landscape is the middle ground of American environmentalism.

This time is different.

It's the cry of politicians, economists, educators, reformers, and scam artists everywhere. We've learned. Today we have better technology. The old-timers were wrong or ignorant, but we know now. Yet most experiments, like most mutations, fail; we don't yet live in a post-ironic culture. The stock market crash of 1929 won't repeat; but market crashes still happen. The overreaction to an assassination in 1914 won't recycle into another ruinous global war; but the overreactions to events in 2001 did. Dazzling new gadgets make communication easier and faster; they don't change how people use them to advance their interests. The numbers add up; the people don't. As literary theorists note, there are only a handful of basic plots, and they date back to antiquity. In fundamental ways the plotlines of character and conflict remain unchanged from Aristotle's *Poetics*. What changes are the particulars by which each generation renders them.

But sometimes circumstances mean that a qualitative phase change, a hydraulic jump in history, does happen. The setting needs only the right spark to propagate into genuine reform. Rarely does such reform produce the outcomes predicted; but change there is since action, any action, forces more action. What lends credence to the belief that the muddling, the reshuffling, or what, after the 2011 season, might be called the wallowing into which the fire revolution sank might this time yield to transformative change is the pervasiveness of the institutional reforms, the sense of alarm that the fires welling out of the West like volcanoes serve no one's interests, and the simple fact of generational succession. The fire officers entering today's ranks were born after the 1988 season or came of age after the National Fire Plan and the era of megafires. They can pick up the torch from where the generation exhausted by fire's culture wars dropped it.

One of the hopeful portents is that the project does not designate the kind of outcome that promotes irony. In that regard "restoration" remains a problematic term—better to call what results "regeneration" or "renewal" so that there is no putative standard in the past against which results can be measured, with the inevitable invitation to ironic sneering. Besides, "historic range of variability" may not mean much in an era of climate change, global species migrations, and a national population that has doubled since the fire revolution began. More hopeful is the return of deep pragmatism, of life conceived as an experiment, once embedded in the national DNA, which seems to be reexpressing itself, like a recessive gene that suddenly reappears.

The blurring of borders—geographical, intellectual, political—has granted room to maneuver. That circumstance is a reprieve, not a permanent condition; it will pass. Nor is 4FRI the only way to reassemble the shards of the fire revolution; even under the Collaborative Forest Landscape Restoration Program a score of other proposals have emerged. Inevitably, new boundaries will be inscribed, and with them will come new rules, frustrations, arguments, and breakdowns. It may be that fire management, like narrative, has only so many plots to choose from and that all its stories will end alike.

Yet perhaps, maybe, this time we are on the cusp of a phase change, one of those rare moments when a precipitating agent can cause a cauldron of chemicals to crystallize out of solution, and for a while we will have some sway over how that happens. No one can say if 4FRI will be the future of fire in the American West; but if it isn't, the continued deleveraging of America's fire economy promises to be a long and painful ordeal, and the country will find that in fire, as in other matters, the fringes will hold only if the middle does.

Coda: 4FRI had two cardinal virtues. It dealt with lands that were contiguous national forests, and thus under a common administration, and it proposed a way to pay for treatments without the direct federal subsidies that had characterized predecessors like the White Mountain Stewardship Project (then costing $500 per acre). Sadly, those virtues apparently turned to vices. Administration fell to the Forest Service's Southwest regional office, which critics believed was still controlled by old-school foresters who bridled over 20 years of timber wars that had shut down logging, and like nearly all foresters they insisted that forestry could pay for itself. The FireScape Project in southern Arizona had no such illusions. But the northern forests had, for a few decades, run big timber operations, a memory that died hard.

The upshot was that the agency issued the thinning contract to a Montana-based company, Pioneer Associates, which promised to convert the small-tree cellulose into furniture wood and biodiesel. The bid was flawed, its proposed technology was quixotic and untested, its managers' previous companies had ended in bankruptcy, and nothing happened on the ground. Rival bidders cried foul and prompted journalists to investigate. In September 2013 the Forest Service announced that Oman-based Good Earth Power was assuming the contract. No details were released.[6]

It speaks poorly for the agency that it handled so ineptly a flagship project that was intended to restore its credibility. The stumble was a deep insult to those within and outside the Forest Service who had labored so long to make 4FRI work. Yet it was, in the end, a delay. In the life of a forest two years is not very long. Still, two fire seasons, 2000 and 2002, were enough to burn away twice as much land as 4FRI proposed to treat over two decades. If the premise was to get ahead of the fire problem, two years, if they were the right two years, might as well be two centuries.

THINKING LIKE A BURNT MOUNTAIN

ON SEPTEMBER 18, 1909, a young Aldo Leopold, then a ranger with the U.S. Forest Service, shot two timber wolves in Arizona's White Mountains. He noted the episode casually in a letter home. But the incident, like embers in an old campfire, glowed in his mind, and in April 1944 he wrote one of his most celebrated meditations, "Thinking Like a Mountain," in which he described standing over the dying she-wolf and watching the "fierce green fire" in her eyes die and wondered if shooting the wolf had helped unhinge the larger landscape. Too much emphasis on safety, he thought, was dangerous. He quoted Thoreau's dictum, "In wildness is the salvation of the world."[1]

The essay, or more accurately moral epistle, became one of the founding documents of 20th-century American environmentalism. It helped make the wolf the living emblem of the wild, and wolf restoration a measure of ecological enlightenment. About 10 miles southeast of Leopold's kill site, Mexican gray wolves were reintroduced in 1998. But his insights also helped underwrite a campaign of nature protection that focused on the preservation of pristine lands. Leopold was the architect of America's first "primitive area," the Gila, located in an adjacent national forest, which subsequently became the inspiration for a National Wilderness Preservation System 40 years later. In 1984 the system acquired the 11,000-acre Bear Wallow Wilderness, about 10 miles as the crow flies southwest from where Leopold shot his wolf. Between them the three sites form a triangle of environmental thinking transformed into action—the deed into an idea, the emblem into a restored species, the wild into a legally gazetted preserve.

A century later a mammoth wildfire boiled out of the Bear Wallow Wilderness, blasted over the wolf reintroduction site, and overran Leopold's vantage point above the Black River. The Wallow fire, kindled by an untended campfire, burned 50 times as much land as the wilderness held. An idealistic green fire met an all-too-real red one.

The contrast almost overflows with symbolism, but two themes seem most useful. One speaks to nature protection, and that preserving the wild is perhaps not just a paradox but an example of a misguided urge toward safety, in this case the security of nature, not unlike Leopold's shooting a wolf. "In those days we never heard of passing up a chance to kill a wolf." Fewer wolves meant more deer, and no wolves meant "a hunter's paradise." So, too, it has seemed self-evident that removing the human presence would mean a healthier land, and no people would mean paradise.

The other theme is fire. At the time Leopold killed the green fire, he was also swatting out red ones. Fire control was the among the most fundamental of ranger tasks; to ignore fire could be cause for dismissal. Interestingly, posters from the era even equated fire with wolves: the fire wolf running wild through reserves was a ravenous killer that needed to be hunted down and shot. Over time this belief, too, yielded to the realization that fire's removal, like the wolf's, could unravel ecosystems. The difference was that fire was renewed annually, if not through human artifice then through lightning. The spark is always there. If wind and fuel are aligned, fire can spread.

But the deeper story was that the sparks decreased and the fuel was stripped away. Lightning fires were attacked and extinguished at their origin. People quit setting tame fires to substitute for nature's wild ones. And overgrazing slow-metabolized on a vast scale what fire had formerly fast-burned. Cattle and sheep cleaned out the country's combustibles. Flame might kindle in the isolated snag; it could not easily spread. Over decades, however, the removal of predatory fire allowed a woody understory to flourish, akin to the metastasizing deer population that blew up after the wolves were extinguished. Both yielded a sick, impoverished landscape.

So a campaign to restore fire ran parallel to that for reinstating wolves. Their histories are oddly symmetrical. The population of neither wolf nor fire has reached its former levels, and the landscape teeters on a metastable ridgeline. The issue is that success requires not merely the presence of wolf and flame but a suitable habitat in which they can thrive. The power of fire resides in the power to propagate, and that sustaining setting was gone. Fire, however, had other properties wolves lacked, notably a capacity not simply to recycle but to transform. A single spark could transmute thousands of acres almost instantaneously.

On Memorial Day weekend, May 2011, flames returned. This time they came as feral fire. It was certainly not a tame fire—not a controlled burn or a prescribed one suitable for wildlands. Neither was it a truly natural fire; it started from a slovenly kept campfire and burned through forests whose structure had for decades been destabilized by logging, by grazing that had destroyed their capacity to carry surface fire, and by doctrines of fire exclusion that had prevented nature's economy from brokering fuel and flame. The Wallow fire could no more behave as it would have in presettlement times than could a wolf pack dropped into a former hunting site now remade into a Phoenix shopping mall.

Probably fires had burned as widely in the past, but through long seasons in which they crept and swept as the mutable comings and goings of local weather allowed. Undoubtedly, in the past spring winds, underwritten by single-digit humidity, had blown flame through the canopies of mixed conifer spruce and fir and left landscapes of white ash and sticks. But it is unlikely that earlier times had witnessed a similar combination of size and intensity. The Wallow burn was not what forest officers had in mind when they sought to reintroduce the ecological alchemy of free-burning flame.

The Wallow fire has not destroyed Leopold's parable, any more than its flames have destroyed the forest in which it took place. But in burning over story and landscape it has transformed them. A burned mountain might think differently than an unburned one. What emerges out of a meditation at the kill site today is a modern parable of the Anthropocene. It describes what has happened when the Earth's keystone species for fire changed how it did business.

The Wallow fire hinges on two paradoxes. One is that industrial societies—those most ravenous of natural resources—are also the ones prone to create nature preserves on a significant scale. The curve of formally protected nature reserves traces closely the curve of fossil-fuel combustion. The other is that a relatively unbridled capitalist society, in the full flush of the Gilded Age, set aside roughly a third of its national estate to shield against the ravages of its own economy. A significant fraction of that (190 million acres) America has committed to national parks and wilderness.

When humanity opted to boost our firepower by burning lithic landscapes, we quit burning living ones, and we used our enhanced fire engines to suppress what fires nature or accident kindled. Apart from technological substitution, the shift was a deliberate policy intended to protect natural places. The founding legislation usually identified sites as sanctuaries "from fire and ax." The immediate outcome was success: forests remained standing and conflagrations halted.

The longer consequences took several decades to become apparent. The shift from external to internal combustion shocked the system as surely as overgrazing or clear-cutting—it just wasn't as visible. The land adjusted in ways that left it out of sync with the kinds of fires the region would experience. When hikers in the Bear Wallow Wilderness left a campfire on May 30, 2011, it did what similar fires had not been able to do for a long time. It blew up and bolted across the length of the White Mountains, overrunning 538,000 acres as it went. The legacy of the past powered that conflagration as surely as the June winds. What happened on the Apache National Forest has been repeated across the American West for nearly 25 years. Attempting to banish open flame has made fire unmanageable, which may be synecdoche for the application of our new, industrial firepower generally.

We now know that attempting to abolish fire from natural sites for which it is indigenous is a mistake as profound as mindlessly extirpating wolves. Yet letting wildfire ramble amid the global metastases of the Anthropocene is an act of faith-based ecology. If all we want is the wild, we will get it. If we expect a usable mix of ecological goods and services, we will have to add our hand to nature's. We created an ecological insurgency, and only controlling the countryside can quell it. Yet to intervene may violate the norms of the wild.

The classic preservationist solution is to leave the landscape to sort itself out, even if this takes decades and the outcome is unlike anything experienced before. Nature has deep powers of recuperation; it has been coping with fire since the first plants appeared on land 420 million years ago. But when wildfires off the evolutionary charts are burning areas 100 times the size of the smallest legal wilderness, there may not be much resilience left. Our nature reserves, even the largest, will burn with properties probably not seen before and on a scale not previously experienced. Paradoxically, their recovery may depend on the character of their surrounding landscapes; these we can tweak into working landscapes to advance biotic goals. Such a conception inverts our traditional notions because it means that the preserved core may not be the refugia for its surroundings but that those surroundings are the source of recovery for the nominal core.

Aldo Leopold's parable was written at a time when the state had to intervene to halt the ravages of global capital and folk migrations. Its primary means were to set lands aside as reserves and to shield them from the practices that wrought wreckage beyond those borders. That strategy kept out landclearing, settlers, and abusive burning. But the internal management of those lands has made them prone to flare-ups and crashes and the losses of what they were established to conserve. In the 21st century the state will likely need to intervene within those reserved lands to prevent the indirect ravages of the Anthropocene.

From the rimrock overlooking the Black River we can see how the Wallow fire integrated everything done and undone over a century with the silent stresses encoded in the name Anthropocene. There is no way to keep out climate change, unhealthy biotas, invasive species, beetle and budworm swarms the size of states, and the relocation of carbon from the Paleozoic to the Colorado Plateau, there to burn again. To do nothing is to risk losing everything save the notion of the wild itself. To do something will guarantee errors. It was easy to identify the causes of land-scalping. It's tricky to track the tremors of the Anthropocene and to know too little from too much safety. The choice is not as clear as whether to pull the trigger or not. And what we might see in the red eyes of a dying fire regime is an exercise in pyromancy. What is clear is that we will likely have lots of occasions to look.

RISING FROM THE ASHES

WHY DOES PHOENIX EXIST? At first blush it seems an odd question. The Phoenix metropolitan area has 4.3 million people and stands as the 14th largest metro area in the United States. Something this big couldn't flourish without cause. It's hard to imagine an urban cluster of this magnitude existing solely for amenities. But if "why" sounds too metaphysical, then try "how."

The traditional explanation is water. Phoenix resides within an archipelago of oases that characterize the arid West. It rests beside the Salt River, which became the site of the first high dam of the Bureau of Reclamation (Roosevelt, opened in 1912). Irrigation agriculture provided an economic basis; those impounded waters could later be repurposed to urban needs. Hydropower furnished electricity. The modern canal system resurrected an ancient network created by the Hohokam before being abandoned, perhaps in the face of megadroughts, in the 14th century. When reborn communities 500 years later revived that network, the name "Phoenix" came easily to mind.

The water scenario holds through the mid-20th century. The population of Phoenix proper was 40,000 in 1940, and of Maricopa County, 186,193. The postwar boom saw roughly 50 percent increases by decade. The 2010 census put the city's population at 1.3 million, and Maricopa's at 3.8; 9 of Arizona's 10 largest cities lie within the Phoenix metro area. Where water and fire met was in the mountain watersheds that flowed into the Salt River. Water rather than timber was the justification behind the forest reserves that claim 20 percent of the state's land. The Central Arizona Project begun in 1973 brought another 1.5 million acre-feet a year from the Colorado River to Phoenix and Tucson. In 1979 the state approved landmark legislation to govern groundwater.

Yet the driver behind growth in the postwar era had shifted from water to fire. Per capita water usage declined as city replaced farmland: for a given amount of water, it was possible to grow 10 acres of tract houses for every acre of flood-irrigated cotton or citrus. What powered urban growth was growth itself, and what powered sprawl was the automobile. Though water remains a thematic obsession—legitimately—the price of gasoline matters more than the price of water, and air quality, not water quality, is the public health menace. What nearly brought the metro region to its knees was not from the rupture of an arterial canal but a broken pipeline in 2003 that transported gasoline from El Paso refineries. The city nearly stalled because, as one wag put it, the valley had become a place of "four million people and five million cars."

The metro area has one of the nation's highest ratios of roads to people. Fossil biomass paves the roads, fossil fuels power the autos that run over them. Coal-fired power plants furnish the power that a built-out hydro network no longer can. Until checked, temporarily, by the Great Recession, sprawl overran land with the abandon of buffel grass. Without its combustion economy the city would shrivel back to alfalfa and creosote. If the per capita consumption of water has dropped, the consumption of combustion has swelled.

If the legacy obsession with water disguises the reality of fire, so internal combustion obscures the nature of that fire economy. Yet its roads are Phoenix's lines of fire. Its high-rises and malls, awash with lights and purring with air conditioning units, are its fields of fire. Its lawns are cut, its trees trimmed, its leaves blown, its people transported, its food and water moved by a menagerie of internal combustion engines. Air conditioners determine the microclimate of habitable spaces; an urban heat island deforms the local climate; the effluents of the new combustion are even unhinging the regional climate. Such firepower sustains the human demographics, which in turn has reshaped its ecology—ornamental plantings and imported pets have multiplied the biodiversity of the land many times over. Industrial combustion waxes and wanes with economic cycles, much as open burning grows and shrinks with climatic rhythms.

The likely check on growth is less water—even rudimentary conservation measures have barely begun—than air. Air quality deterioration from the ceaseless belching from auto tailpipes eats at the quality of life in ways that ripple widely. The scene, in brief, is an industrial firescape.

How that firescape interacts with the firescapes of open burning says much about the future of fire in the Southwest. The two combustion economies tend to compete rather than complement.

The fire-water linkage endures. The various factors that are promoting larger and more severe fires—particularly climate change, proliferating exurbs, and a reliance on mechanized suppression, all of which derive from industrial combustion writ large or small—are unpicking the watersheds that still bring water and power to the valley. The Salt River Project is expending money and political capital to spare its watersheds from scouring megafires. The methods proposed for remediation all rely on political actions, which will be decided by urban voters. Few of those voters know fire personally other than what they see on TV or LCD monitors. They no longer burn off their lawns or leaves. They find open flame, even candles, banned from residences. They can't light up cigarettes indoors in public places. And they find it increasingly tricky to burn ceremonial fireplaces because of no-burn days.

The reason points to a shift in attention from watersheds to airsheds. The winter days that are cold enough to warrant a fire are precisely those characterized by inversions that trap particulates and combustion gases and so fall under no-burn restrictions. For a place that, because of internal combustion embedded in cars and trucks, teeters on the cusp of air quality violations, the injection of wood smoke is a cost it cannot afford. As the constraints become more cumbersome, it is easier to eliminate burning than to try to regulate it. In 1975 the smoke from broadcast burning in the pineries of Fort Apache followed the Salt River gorges and socked in the valley. In 1996 the smoke from the Lone fire on Four Peaks threatened to close Sky Harbor Airport. There is scant tolerance for such events. The firescape has been remade according to industrial combustion: the community will accept effluent it can't see but not that it can.

By flinging out new exurbs, and by remaking old rural communities like Prescott and Payson into outliers of metro Phoenix, these urban values extend widely. The valley may occupy only 1 percent of the state's landmass, but the same might be said of a black hole in a galaxy. Its gravitational influence extends throughout most of the state, and reaches far wider through the establishment of de facto fire protectorates, as each exurb declares its own zone of concern and erects an airshed canopy, and as it wields its political influence. The keystone species for fire remains humanity.

More and more, the decisions that matter are made in metropolitan areas, and in Arizona that means Phoenix. It is indeed rising from a fire, but its ashes are the soot of industrial combustion.

UNDER THE TONTO RIM

"You mean—"
"Mebbe we can't get out. The forest's dry as powder, an' thet's the worst wind we could have. These cañon-draws suck in the wind, an' fire will race up them fast as a hoss can run . . ."
"Good God, man! What'll we do?"

—ZANE GREY, *THE YOUNG FORESTER* (1910)

LOOK NORTHEAST FROM PAYSON and see the scar, still livid after more than two decades, of the 1990 Dude Creek fire that ran under and along the Mogollon Rim, flaring through the Bonita Creek and Ellison Creek subdivisions and burning over a fire crew of inmates from the Arizona Department of Corrections. Sixty homes were lost; five inmates and one guard died. The driving force was a microburst from a collapsing pyrocumulus. The scour line, yet raw, looks over the Payson valley and its tumbling exurbs like a corroded tombstone. Look northwest and you see nothing but rumpled pine and rimrock. The problem here is what might happen, because the old villages of Pine and Strawberry, miniatures of Payson, lie in a trough below the rim. The Dude Creek scar reminds the fire community what it can cost to protect an exurb amid so volatile an environment. The unburned but at-risk villages remind it what the cost of doing nothing can be.[1]

The Tonto Basin stretches like a vast, sagging hammock between the Mogollon Rim and the Mazatzal Mountains. It's higher and wetter to the north than to the south, and in the east over the west. It's a rumpled landscape below the rim, with a few broad shallow valleys, of which the Pleasant and the Green are best known. Until the late 1870s it was Apache country—the local band was known as the Tontos. When that menace ended, the scene flooded with livestock. The early accounts speak of waving bunch grass stirrup high, though that had to mean a wet year. For pastoralists it was a land of milk and honey; fodder was abundant, springs bubbled, and creeks flowed. In 1882, roughly coincident with the last battle between renegade Apaches and the U.S. Army (not far from Dude Creek), Payson was platted. Two years later it began an annual rodeo—the oldest continuous one in America.

It didn't take long for land and people to sour. The herders quarreled, climaxing in the long, lethal feud between the Grahams and Tewkesberrys known as the Pleasant Valley war. But as the vendetta wore on, there was less to fight about. The trampling, the throngs of livestock, the staccato wave of droughts—over the next three decades the grass wore out, the springs dried up, and the cattle were sold off or died. Meanwhile local saw mills were stripping away the giant yellow pine. In wet years with a good seed crop pine reproduction flourished, most spectacularly a 1919 irruption, which did for pine what predator destruction did for the Kaibab deer herd (without fire there was nothing to prune back the brush and dog-hair thickets). By the time Depression slammed the Tonto Basin the Old Forest was being replaced by a self-strangling tangle of manzanita, scrub, oak, juniper, and thronging pine—a briar patch of brush and stunted poles, less a forest (much less a forest savanna carpeted with bunch grass) than a conifer jungle. The surface was as littered with biotic shards as with ceramic ones. As Aldo Leopold noted, Arizona's brush was different from California's, not the outcome of climate and soil so much as of abusive grazing.[2]

All times, all places, all biomes—everything everywhere is always in transition, its species always mixing, its stories stirring together. But some times and scenes are more sharply transitional than others. Geologically, the Mogollon Rim traces the abrupt edge of the Colorado Plateau as it tumbles down a series of tiers into the Basin Range. Biologically, the land under the rim cascades from montane forest to Sonoran desert, with plenty of relic and pioneering species jumbled together in between, as the biomes seem to breathe, advancing and retreating, pausing and staying, pulsing with the climatic bellows of wet and dry decades. It's a chunky patch quilt of ecological communities, with a latent, hair-trigger understory ready to thicken into any of a variety of forms. Even in terms of human history, the story can veer suddenly into any of several paths. The replacement of Apaches with Anglo herders was one such moment. The replacement of ranchers with recreationists was another. Until the Beeline Highway between Phoenix and Payson was paved, however, this was a remarkably isolated land. Its contact with the outside world came from migratory herds and flocks and from the words written about it. What happened under the Mogollon Rim stayed under the rim.

Yet even as the land dry-rotted, the landscape became famous as an epitome of the Old West. In 1906 Zane Grey first trekked to Arizona; the next year he visited the Grand Canyon's North Rim. What he found, and what he would celebrate in a volcanic pile of pulp novels and movies was freedom, abundance, and adventure. His output flooded the best-seller lists from 1917 to 1924; by then he was turning pulp into celluloid as well. Between 1918 and 1929 Hollywood filmed some 32 silent movies based on his Arizona romances (113 based on his whole oeuvre). Not until the Depression and a stroke did Zane Grey Inc. slacken its production. Between them his novels and the films, liberally peppered with essays for the slick-paper magazines, helped gel the American conception of what the Southwest looked like and what it meant.[3]

Grey ranged widely, and he wrote about all the main provinces of the state from the Grand Canyon to the Yuman desert, but when in 1920 he decided to build a hunting cabin, he erected it outside Payson. It lay under the rim, looking out onto the timber-terraced valleys below. The Mogollon, named for Ignacio Florés Mogollón, a governor of New Mexico for New Spain, Grey decided was too difficult for Anglos to pronounce, so he borrowed the name of the local Apache band, and recast it as the Tonto Rim, which is how is best known today. Under the rim's looming facade, he wrote a stream of books.

Unsurprisingly, a few featured fires. An early romance in Arizona, *The Young Forester*, appeared in that national Year of the Fires, 1910, complete with harrowing flight in which the hero is saved from monstrous flames only through a daring backfire. Fire, the chapter title shrieked, was "The Forest's Greatest Foe." In another, *Wildfire*, horses and flames compete in a deadly race. Fire was a part of the scene, like bears and cougars, and coping with it was a frontier skill, as familiar to rangers as a diamond hitch. The Payson cabin remained Grey's base of operations.

But what he saw from its porch was changing. His old guide, Al Doyle, died, and Doyle's son, Lee, became a tourist's guide and owner of a movie-star horse named Rex, "King of the Wild Horses." Had he chosen to be more reflective, Grey might have seen that the same was true for himself. He no longer lived the Old West: he wrote about it and filmed it. Nor was it simply the characters of the Old West that had passed on. The Pleasant Valley landscape had changed as surely as if it had been plowed under or paved over. In 1930 Grey argued with the state game commission about hunting a problem bear out of season. When refused a license, he publicly declared he would never return to Arizona. Why? The forests and ranges were overrun and scalped, tourists had replaced working cowboys and hunters, the wild was tarnished and tatty. (Aldo Leopold had a few years earlier vowed not to return to the Southwest of his youth after gravel roads replaced horses.)[4]

Of course Grey declined to complete that syllogism: that it was the pioneer settlers he celebrated who had promoted the logging and herding, that his books had become Baedekers for well-heeled tourists in search of the Old West. But he was right about the landscape. The lands under the looming rim were a shadow of their former selves. By the 1920s the rim country had become a miniature of the self-destructing West generally: the grass eaten out, the creeks shriveled, brush and dog-hair conifer thickets replacing clumps of fescue and muhly, the game gone, the big trees hauled off, the indigenes diminished and on reservations, with the spoilers feuding among themselves when they weren't lamenting the lost past and promoting that vanished past to tourists.

And one should add to this litany of abuses: the fires were knocked flat. A rhythm of burning, an ancient fugue between climate and people, broke almost instantly. The last regional-scale fires cease almost precisely as cattle and sheep swarmed over the high valleys. Again, Aldo Leopold put the matter pithily in 1924, observing that "when the cattle came, the grass went, the fires diminished, and erosion began." Watersheds that had survived hundreds of years of burning collapsed from a handful of years of overgrazing. What "took over the country" was a woody understory collectively dismissed as brush. (Curiously, that same year Grey wrote "A Warning to California" in *Outdoor America* about that state's spiraling fire scene.) Payson Ranger Fred Coxen estimated that as many as 1.5 to 2 million head of cattle had descended on the range with the suddenness of a hailstorm. By the late 1920s he remarked that the Tonto Basin had reached "the ragged end of it all." Then Depression tag-teamed with drought. Zane Grey decamped just as that defraying end tipped into apocalypse. A famous photo from 1931 showed not mountain lions treed by baying hounds but an emaciated steer lying dead on soil stripped to dust outside a fenced exclosure rife with perennial grasses.[5]

The Payson cabin fell into decay.

For the rim country around Payson revival came in a postwar boom driven, literally, by the automobile. Before then transport was mainly by wagon or, especially, pack train. Livestock moved into and out of the upland valleys in herds; sheep were still tramping up and down from desert to rim even after the war ended. The paved Beeline Highway reached Payson in 1958, realigning the economy of the rural scene. The "one man, one vote" rulings of the Supreme Court from 1962 to 1964 remade Arizona politics, shifting power away from ranchers toward suburbanites.

Payson's human population exploded, as its population of cattle had the century before. In 1912 the town had 125 people; in 1930, when Grey decamped, it had 500; and two years after an asphalted Highway 87 reached the town, 814. By 1980 it had supernovaed to 5,068; by 2000, 13,620; and as the economy tipped into the Great Recession, nearly 20,000, before the services and amenities economy crashed and burned and the population sank to 15,301 in 2010. And these were full-time residents recorded in the census; no one knew the tides of people that ebbed into and out of the land. In effect, seasonal tourism had replaced the transhumance of livestock. A swarm of unregulated overbuilding remade the landscape as overgrazing had previously.

Payson was soon reincarnated into an exurb of the Southwest's largest metropolis. No comparable road connected Pleasant Valley as the Beeline Highway did Green Valley, so it was Payson, not Young, that became the entrepôt below the rim. The rim's big exurb began breeding its own echoing exurbs, reclaiming old rancher homesteads for second homes and summer camps. Paving later extended to Pine and Strawberry, more diminutive versions of Payson— villages, really. Still smaller versions, second-home subdivisions, littered the countryside like abandoned mining claims, some 26 orbiting around Payson proper and 67 in the Payson Ranger District. Much as early ranchers had homesteaded small plats near water and left their stock to graze freely on public lands, so new homesteaders reclaimed those pockets of private holdings, living off the amenities of the public lands around them and leaving fire protection to agencies of the state. In 1966, as if on cue, the Grey cabin was rebuilt as a tourist site.

It was happening everywhere: from 1950 to 2010 the population of the United States had more than doubled, and an ever-restless people were recolonizing former rural lands with an urban outmigration. Exurban sprawl was penetrating into every nook and cranny of the post–World War II era as the cattle and sheep had swarmed over western rangelands in the post–Civil War era; by 2010 the Tonto recorded 5.8 million visitors a year at some 1,600 designated recreational sites. The new Gilded Age often favored log mansions in the woods rather than estates overlooking Connecticut beaches. Why did they favor the seeming wild? There are many reasons, an environmental movement among them, but also residual sentiments about what the West meant, a sentiment left like stumps and sprouting Gambel oak from earlier eras. The Tonto's cultural landscape had an understory of literature, art, and folklore as ready to spring into prominence as its brush if the political overstory opened up.

If there seemed a disconnect between what people had in their minds and what the land actually showed, that too should have surprised no one. The newcomers could celebrate the sense of freedom Grey had championed, though of a different sort, an open range of second-home pioneering, and they could relish the great green forest, even if they were more likely to bushwhack through clumps of scrub than gallop among big pines like Grey's western cavaliers. Still, for those unfamiliar with the Old Woods, or who read the landscape through Grey's purple-sage prose, it seemed lush and forested. Viewed from above, a green sea of swelling conifers appeared to lap against the storm-tossed cliffs of the rim. Newcomers accepted the landscape of their youth or first encounter as the norm. They viewed with eyes that saw what they wanted to see.

Fires were plentiful. The transition zone between the Colorado Plateau and the Basin Range, slashing across Arizona from Prescott to the White Mountains, is prime territory for lightning fire. It receives enough moisture to grow grass, brush, and pine, but not enough for shade-tolerant mixed conifer; enough to hurl thunderheads up but not always enough for rain, which could melt away into virga. Dry lightning is common during the early summer monsoons. Over desert the downdrafts can stir up dust storms; over flames, they fan firestorms. How much the Apaches added to the regional fires is debated. On his 1900 tour Gifford Pinchot described how "from a high point" on the rim a bit east by the White Mountains his party "looked down and across the forest to the plain. An Apache was getting ready to hunt deer. And he was setting the woods on fire because the hunter has a better chance under cover of the smoke."[6]

The Tonto National Forest, which spreads from the rim to the desert Superstition Mountains and the brushy Mazatzal Mountains, is by nature a fire forest that has also become by proximity to Phoenix an urban forest. It's a big reserve, the fifth largest in the national forest system. Its fire load places it among the top 10 percent; the Payson District alone averages over a hundred starts annually.

The trampled landscape helped keep fires in check: the grasses that had allowed flames to creep and sweep below the rim were largely gone. After it became a national forest in 1905, the Forest Service installed formal fire protection. The CCC built an infrastructure, including a fire control road and fuelbreak that ran below the rim from Kohl's Ranch to Pine, a miniature Ponderosa Way. During the late 1950s the Tonto had experimented with controlled burning to improve range and watershed, but that passed. Without a continuous cover of grass, fire could be tricky to control; as a policy, prescribed fire was suspect; and too often smoke drifted down the Verde River to the Salt, and down the Salt to Phoenix, whose residents found fire neither a valued tool nor an adventure. Instead, the Tonto advertised high-impact fire protection.

Meanwhile the scene intensified, its parts tightening like a coiled mainspring. The woods thickened and stagnated, stocking the landscape with more and coarser fuels; houses sprouted like juniper on the patches of private land; the fires kept coming. Eventually, the separate pieces would fall into place, like tumblers in a lock, and this time there would not be the extensive buffer lands that a daring ranger could use to backfire. The convergence to catastrophe happened on June 25, 1990, when dry lightning kindled a fire near Dude Creek, under the rim northeast of Payson, amid the hottest day of record for Phoenix (122°F). What followed was a chronicle of lethal transitions piling one on another.

The fire blew past initial attack. It ran unchecked while a Type II team was assembled. It burned vigorously through the night while the fire team tried to transition to a Type I team. The next afternoon, June 26, thunderheads boiled up from the rim, like spume from storm waves crashing into headlands. By 1 p.m. a massive plume had developed over the Dude fire and by 2:30 p.m. it was collapsing and shot downdrafts that sent the flames racing a mile and a half in an hour, scouring out the rim like the cutbank of a river in flood. Half a dozen crews were strung out in Walk Moore Creek, preparing to burn out to protect the Bonita Creek Estates. Alert crew superintendents recognized the warning signs and readied a safety zone with bulldozers to the north. The two crews farthest out, the Navajo Scouts and the Perryville inmate crew, fled south. The Scouts made it to the Control Road (FR 64) and were driven away in vehicles. The Perryville crew was split by flames running like a boar's-head wedge through a Viking shield wall. The northern squad, cut off, tried to deploy shelters. A few survived unscathed, five survived with burns, and six died.[7]

The Dude fire tragedy was a dress rehearsal for the South Canyon disaster four years later outside Glenwood Springs, Colorado, and an eerie premonition for the Yarnell Hill tragedy in 2013, which followed a similar scenario within the same transition zone; the one stands to the other as the Savings and Loan scandal does to the Great Recession. Had that lightning-kindled fire been farther east, it might have overrun Young. Had it been farther west, it might have blasted into Payson. Had it been around the bend of Milk Ranch Point on the rim, it could have surged over Pine and Strawberry with losses far greater than the 60 trailers and cabins at Bonita Creek Estates.

There was rich symbolism when the Dude fire burned the rebuilt Grey cabin. The imagined world penned on its porch was gone; first from its own careless disregard, and then by flame. That scene was a cliché, a pyrogeography written by a pulp-fiction populist, though that didn't make it less real. The rebuilt West, equally careless in its own way, was becoming as clichéd as its exurbs. What made this site special was that it played out with a Zane Grey title hanging over it. Living under the Tonto Rim meant living under the shadow of fire.

It's easy to trash a landscape, hard to repair one. On the Tonto it began with reducing livestock. In 1910, the year of the Great Fires in the Northern Rockies, the Supreme Court agreed that the Forest Service could regulate grazing on the national forests. Still, it was local nature not political coercion that drove the numbers down. Its grasses scarce, its droughts recurring—the Tonto's landscape could do what legal threats could not. By the time of the Dude fire one cow grazed where a century before a hundred had. But until woods and brush were restructured, creeks would be a trickle of their former selves, and with the right timing fire could wash over whole landscapes like the downburst from a monsoon thunderstorm. It wasn't possible to rebuild the Green or Pleasant valleys to their past glory. But it might be possible to shield Payson, Pine, and Strawberry from the fate of Bonita Creek Estates.

The process began after the ashes of the Dude fire had settled. In the mid-1990s fire officers turned to fuelbreaks—the fences of open-range fire—around Pine. The community was nestled in a deep side canyon of the rim; a box canyon, really. A fire could push through it, or more likely millions of embers, a blizzard of sparks, could rain down on it from the rim. The project was too timid and it faltered for lack of funds and scale. By then injunctions from lawsuits over the Mexican spotted owl shut down what remained of a timber industry, which left forest thinning without a market. Instead, the Tonto's attention turned to invasive grasses in the desert and the fires they powered, which threatened to replace the state's signature saguaros with *Bromus*. In 1996 an abandoned campfire burned Four Peaks in the Mazatzal Range from the summit to the desert, 60,000 acres in all.

What galvanized a new response was the National Fire Plan. What kept it on track was the personal memory of people like Don Nunley. He had been involved with fire on the Tonto since 1985, as a helitacker and then as a member of the Payson Hotshots (on the Dude fire), a fuels specialist after the National Fire Plan, and now the Payson District fire management officer. Patiently, the Payson fire staff turned their attention to the archipelago of in-holdings, some no more than old 40-acre homesteads patented in the 19th century. They began cutting fuelbreaks 10 chains wide around them; and they helped local fire districts build capacity through federal grants. As megafires seized the national attention, and as the Dude scar, like a colossal billboard, advertised the threat, the public began to rally.

When the first-entry fuelbreaks were done, the program pushed out another 10-chain swath where possible. Within a decade the mechanical treatments had roughly achieved their goals: every community was surrounded by a moat of reduced fuels. Already the program was sliding into maintenance mode because treatment was not a one-off project like a vaccination but a process that needed routine boosters. Opening the woods often opened the scene to explosive brush, which meant continuing care, which meant more cutting, on a three- to five-year cycle. The lumpy terrain and loose soils argued against machines in favor of more expensive hand labor, and the dappled geography of in-holdings argued against broadcast burning in favor of piles. In August 2014 the Payson fire staff estimated they had 75,000 piles ready to burn over fall and winter. The decade's rehabilitation had cost $16 million, a big sum for a southwestern fuels program but a pittance compared with the megabuck cost of megafires or the fuels programs around Tahoe or Lake Arrowhead.

It had taken decades to recover from the livestock inundation of the 19th century. It would take decades to fashion a more habitable landscape from its successor swarm of second-homesteaders. Nor was protecting exurban enclaves the same as managing the landscape. Some large patches (notably, Mount Ord) got treated, and in the lower elevations big wildfires put fire back in the scene. But in the forest proper there was little opportunity to let fire free-range. The in-holdings are not large in area, either individually or in aggregate, but each demands its own penumbral zone of protection from fire and smoke, which leaves little land available in which to maneuver. The only strategy is a hard initial attack, the hope that treatments will reduce the number of starts that survive long enough to demand extended suppression, and the expectation that escaped wildfires may do what prescribed burns can't, provided fire officers can keep those flames out of Tonto Village, Whispering Pines, Geronimo Estates, and threescore others.

It's a pyrogeography of edges for which there are few economies of scale possible. Transitional places—and the Tonto piles transition upon transition—are landscapes of borders. Their suppression surfaces, as it were, are large compared to the area protected. Purchase is difficult for managed wildland fires because as soon as a project gains traction, it runs into another tiny patch from which fire must be excluded. Terrain—a geology of hills cascading from the rim—aggravates the problem by limiting mechanized equipment and requiring hand labor. The legal, political, and economic costs escalate.

So, over the past 50 years, the fire scene has resembled the open-range scenario in which the bias favored the herders: if you didn't want their cattle in your garden, you had to fence them out. The open range died hard. But an open-range philosophy rides on, the default setting for American politics and economy. If you want privacy on the Internet, it's your job to fence off your identity. If you want to run herds of houses, you can. Regulating the social costs of private profits is a hard sell; over its history, the United States has preferred to pick up the mess afterwards. Today, there are few opportunities to rein in the building boom. The private costs, however, are beginning to eat into public purses as each summer's toll of burned communities mounts.

The move toward a landscape scale, however, slogs along. The Tonto has joined the Four Forests Restoration Initiative, though its timber contribution is minor (1,000 acres a year). More significant is NEPA approval for five analysis areas over the district as a whole, which lubricates away much of the social friction. Yet even as the program begins raising sea walls around every isle, the ocean of potential fire is rising. At some point it will again storm over the barriers and we will learn if the walls are enough to quell the surge.

The Tonto has a story to tell. A first reconnaissance suggests it's a WUI story, about a service and amenities economy that colonized a broken landscape with retirees and second-homers and so fragmented habitats and complicated the economies of scale that cost-effective fire management seems impossible, yet somehow is happening. But it's also, maybe more, a story of the West, and of the ideas people have of it. Why are they coming? What do they think they are moving to?

There are, as always, many reasons, some intensely personal. But surely a part is what Grey waxed so zealously about in novels and movies, a sense of elbow room amid inspiring scenes. That freedom may be less one of opportunity than of unburdening from the cares of urban life, while axe-and-hoof pioneering is replaced by nature appreciation. To the ecologically sensitive, the woods may be a shambles, but they are more natural than whatever city the newcomers are fleeing. What is natural is what they first saw. They see through a Grey-tinted glass, if secondarily refracted through a prism of environmental enthusiasms. Any change is for the worse. They don't want to be told what to do and they don't want the woods altered.

Yet the fear of fire is shifting that perception. Like the rain that falls on the just and the unjust alike, the West's big fires are taking everything as they scour away the beneficent along with the toxic. In the Old West, settlers, eventually even ranchers, recognized they could no longer tolerate the trampling of the commons or have vendettas take the place of law. So, today, the settlers of the New West understand the need for some degree of collective rule. Private and public efforts mingle along the edges. Public grants shield enclaves and help build up fire protection districts. Volunteer fire departments work with federal fire crews. Some private communities have even raised money to pay for landscaping on the national forest around them. Since 2006 those sums total $751,110. The Gila County Board of Supervisors authorized $200,000 for fuelbreaks on national forest land around Pine and Strawberry, Payson, Star Valley, and East Verde Park. The East Verde Park homeowners' association and fire district contributed $100,000. Payson anted up another $50,000. Christopher Creek and Hunter Creek raised $88,000. The Tonto Apache Tribe, the Rim Club, Terra-Payson 40 LLC, Pine-Strawberry Fire Wise, and Pine-Strawberry Fuels Reduction Inc. have all added to the pot, like communities in floodplains raising levees.[8]

The economic engine of the Tonto Basin is Payson, but what powers Payson's economy? It isn't the commodity booms of the Old West celebrated in its annual rodeo. The Great Recession showed it won't be the New West of exurban construction or seasonal tourism, which crashed and (nearly) burned. City planners instead place their latest hopes in a new satellite campus that Arizona State University may build on 250 acres of public lands adjacent to the Payson District ranger station. That shift from rodeo arena to lecture hall speaks volumes about the economic realities of contemporary western life.

Change has come, and for an index turn again to Zane Grey. New Westerners are less inclined to read him—thrift stores overflow with matched sets of his books, fifty shades of Grey unable to find readers. Adventure romance is more likely to run to political thrillers or science fiction; Hollywood westerns, to *Brokeback Mountain* rather than *To the Last Man*. Grey is less an oracle than a local-color character suitable for commercial enhancement.

So when his cabin was again rebuilt in 2005, it was relocated to Green Valley Park within the Payson city limits as a museum administered by the Northern Gila County Historical Society. It more resembles a McDonald's Playhouse than a hunting shack. It's not on the frontier: it's not even amid the exurbs. It's a monument, more statue than living museum. But it is less likely to decay or to burn down, and it speaks to a history that is well past. What memories of the Old West—or a fantasized Old West—it symbolizes remain within. It no longer lies under the shadows cast by the rim or the rim's potentially lethal pyrocumulus.

SQUARING THE TRIANGLE

Fire at San Carlos

ON APRIL 19, 2014, a small storm cell passed over the Nantanes Plateau, shedding virga and dry lightning. The next day a fire was sighted among grass and shrubs above Skunk Creek. Two days later smoke spindled up through the pine forest around Bloody Basin. When the fire and forestry staff of San Carlos gathered, the Skunk fire had scooted to 60 acres and the Basin fire had barely held beyond a snag and scorched ground.[1]

The group gathered around a table blanketed with two detailed topographic maps, exchanging confidences and opinions like a covey of physicians over a problem patient. Where, Duane asked, do we draw the box? Until recently such a question would never have been asked: response to a reported smoke would have been automatic. Both fires would already be out by now, or nearly so, as engines, hand crews, and maybe aircraft attacked it. That was still a live option. Now, Duane, Dee, Clark, Kelly, Bob, Dan, Bear, and later, Nate, poked and pointed and passed questions.

The San Carlos fire cadre viewed the reported smokes less as problems or threats than as complicated opportunities. Though larger, the Skunk held lesser interest. A high-pressure ridge had rolled through since the fire began, raising temperatures and stirring blustery winds. Were the fire primed to burn, it would have reached a thousand acres instead of 60. It was, as its name implied, just skunkin' around and offered fewer options to manage. The Basin fire burned, if unsteadily, in country that badly needed burning. Whether this fire, at this time, was the right fire for that job occupied the conversation.

They discussed quietly, each offering thoughts in turn, pointing to the map, pausing often, a conversation of silences as much as of utterances, tracing terrain, searching for consensus, not yet ready to decide a course of action, trying to match the fire with landscape features not documented on maps and with purposes not coded into dispatch software, probing like the flames of the Basin fire as it desperately foraged for fuel. Where might they "draw the box"? They studied old roads, trails, rims and ridges, natural fuelbreaks of any kind. Any roads would have to be improved since lava flakes and basalt cobbles would shred engine tires. They would need to move "heavy iron" in the form of bulldozers and graders to do it, but that too was an opportunity because it would allow them to upgrade their infrastructure even as they reined in the fire. Bear recalled an abandoned trail. Dee lamented that the ideal solution might be the old military road that cut through the Nantac Rim. They traced a contour that, if joined by a handline, could tie two existing fragments together; for that they could use a shot crew, specifically the Geronimo Hotshots, San Carlos's own.

They proposed, in brief, to adapt policy options that allowed fire officers to confine and contain wildfires. They were neither direct-attacking the fires nor leaving them to burn, but loose-herding, boxing-in and burning-out wildfire, and by doing so they were devising a surrogate for prescribed fire, much as prescribed fire had been promoted as a surrogate for natural fire. In this they were part of a trend throughout the West. The policy was latent in the 1995 federal common policy for wildland fire or for that matter with reforms that dated back to 1968 for the National Park Service and 1978 for the Forest Service. What made San Carlos interesting is that they were applying those permitted principles on a serious scale on the ground. Still, some tension was in the air. The best way to prevent a fire from going bad was to hit it when it started, and operations were instinctively geared to act, not to ponder.

They decided they needed more intel. Two crews were dispatched to each fire to document, locate, take measurements, order a spot weather forecast, and assess the potential for spread. A grader was noted, parked not too far from the Basin fire. A bulldozer, now in the shop, would be dispatched before the morning was out. Another weather system, more powerful, was forecast to rumble through in another day, perhaps with rain, certainly with strong winds and a drop in temperature. It would determine if the Skunk could do more than scuttle through scrub, and if the Basin could burn through the forecast evening frosts. Today's intel and the coming front would decide. For now it was enough to confine and contain.

If the fires survived, they would move from statistics to operations. They would go from a threat to an opportunity, from sand gagging the gears to a lubricant. Their management became interesting in the same way that radio-tracking a bear is more interesting than killing it. Instead of automatically responding to chance ignitions, fire officers could manage by choice. That Easter Sunday weekend the resurrection of free-burning fire on the San Carlos was underway.

The San Carlos Apache Reservation sprawls, crudely gerrymandered, over 1.8 million terraced acres and many millennia of terraced time.[2]

Geologically, it presents a series of landscape steps crusted and filled with volcanic outflows. Each terrace replicates and lowers the one above, until the coherence of the rims and plateaus fragments into isolated peaks and valleys. The highest terrace is the Colorado Plateau in the northeast, edged by the Mogollon Rim; its lowest point, the sky islands and valleys of the Basin and Range to the southwest. The highest terraces are flat and filled with basalt. The lowest valleys are stony pediments, sinking into the floodplain of the Gila River.

Ecologically, the terracing appears as roughly contoured life zones. The Mogollon Rim boasts a robust ponderosa pine forest, grading into mixed conifer. The middle terrains are mixtures of woods and grasses; the basaltic terraces range from prairie to high-desert grasslands. The lowest landscapes are Sonoran desert, degraded into creosote. About a third is woodland; a quarter, grassland; a fifth, desert; some 13 percent, ponderosa; and the tiny remainder, human communities of some kind.

The heart of San Carlos is the Nantac Rim, a miniature of the Mogollon Rim, slashing through the middle of the reservation from northwest to southeast. Along it, ponderosa pine flourish, while the terraces that flank it are grasslands—Big Prairie to the north and Antelope and Ash flats to the south. Near its center lies Point of Pines, where a stringer of ponderosa reaches into Big Prairie. It rests at a kind of eco-librium midpoint for rock, biota, and human history.

The human history, too, appears layered, terrace by excavated terrace. In part, this simply reflects the source of the evidence from archaeological excavations, digging down through layered culture upon culture. Instead of cascading across space, eras pile one on top of the other. But people have actively made terraces as well. Ancient agriculturalists terraced hillsides to hold soil and water, 20th-century mining terraced whole mountains to strip off ore, and archaeologists reversing the process have pulled back layer after layer by terraced pits.

Yet, a great fact of the reservation is that it runs cross-grained to that texture of time and space. Drive from San Carlos, the tribal headquarters near where the San Carlos River joins the Gila, to Point of Pines, roughly at its center, and you pass through Sonoran lowlands up through the foothill flank of the Gila Mountains to the Antelope and Ash flatlands, rich with high-desert grasslands, and then through Barlow Pass over the pine-clad Nantac Rim and onto the sweeping plains of Big Prairie. Continue from Point of Pines to Malay Gap, and you rise through a lesser plain, clothed with juniper savanna, and up to the Mogollon Rim itself. Unless you follow the Gila River, to move around the reservation is to step up and down terraces of some sort.

So it is with San Carlos history. It moves in jumps and over barriers rather than along worn pathways or the meandering floodplains of mainstream narrative. Peoples come and go, sometimes abruptly, the strata of eras piled one on another. But as they cross the grain of place, they can pick up quirky and unexpected adaptations that can evolve into telling traits. That cross-texturing of land and history has created nooks and niches with unexpected outcomes that can hold larger, even national significance. This has happened more than once in the past. And it seems to be happening to its wildland fire program today.

The setting is classically southwestern. That means the scenery is broken with peaks, plateaus, valleys, and gorges, amid an arid to semiarid climate that leaves its rocky exoskeleton starkly visible. It means its history is full of sharp human entries and exits. And it means there is plenty of fire.

The broken, terraced terrain; the mixed, middle biomes between valley desert and mountain forest; the annual rhythm of spring drought followed by summer monsoon—all are ideal for sparking fires, frequently by dry thunderstorms. San Carlos averages roughly a hundred such fires annually, with some years showing much less and some far more, and it stands close to the epicenter for lightning fire in North America. If it were a pond and you tossed a stone at Nantac Rim, the resulting ripples would align roughly with the density isolines for naturally ignited forest fires. Like the landscape, its fires burn patchily. In the past, the land burned from the sheer numbers of starts, and when dry years followed several wet ones, from lingering fires that could creep and sweep over many weeks. Only central Florida approaches that intensity of natural burning.

The record of fire is continuous; the chronicle of people, less so; and how they interacted, almost unknown. The terraced history exposed by archaeology begins with an Archaic culture of big-game hunters and foragers. Then, corresponding roughly with the rise and fall of the Roman empire, a Mogollon culture appears and continues until the mid-15th century. The Mogollon exchanged not only goods but styles of pottery and housing, and no doubt other ideas, with the desert Hohokam to the west and the cliff-dwelling Anasazi to the north. The Anasazis began migrating into the region during the 10th century; by the 11th, as Europe commenced its cycle of crusading, they were dominant, and clusters of pit houses became suburbs to Anasazi complexes. Then, suddenly, they all—Anasazi and Mogollon both—vanished during the 15th century. When the westernmost bands of Apache began filtering through the landscape, sieving through mountains and mesas from the Great Plains, perhaps a century before the first Spanish entrada, they passed through a landscape of ghosts. The chroniclers of Francisco Vasquez de Coronado in 1540 called the land a *tierra desplobada*, a place abandoned, a land emptied of people. The small bands of Apache hunter-foragers still dribbling into the land seemed to melt into the scenery.

For a millenium, however, from 400 AD to 1450 AD, the human population had been sufficiently dense to carve agricultural terraces and erect stone pueblos. How those cultures affected the prevailing fire regime is unknown: the earliest fire-scarred trees come later. But deal with fire they had to, because it was all around them. When they left, whatever lines and fields of fire they established fell into ruin as fully as their kivas and apartment houses. Until the Apache established themselves, the land went feral, and the abundant sparks of nature reclaimed and remade the scene. Until Europeans arrived, or more significantly, until the modern era that followed Mexican independence and then the Mexican War that ceded the region to the United States, the characteristic fire regimes were those negotiated between the light-on-the-land Apache and the heavy rhythms of lightning.

Ethnographers and ethnobotanists are adamant that the Apaches did not burn widely: they didn't need to. What they wanted from the habitat—edible wild grasses, game animals, shrubs for baskets, many dependent on routine burning—they got freely from a land drenched with flame. They burned some horticultural plots for special crops, they burned patches (some large) to attract preferred fauna by greened-up pasture and fly-retarding smoke, they burned during raids to alarm enemies and to cover tracks, and they undoubtedly left fire as litter. But human numbers were small, fire numbers large. The Mogollon peoples had sculpted surface lava into stone villages; the Apache assembled shrubs and twigs into wikiups; and so with their fire regimes. There are plenty of peoples of comparable technology—many Aboriginal bands in Australia, the congery of tribes in California—who burned both meticulously and extensively. The existing evidence suggests the western Apache did not.[3]

Their primary contribution to the region's fire regimes was not to add ignition so much as to remove barriers to fire's free propagation. Specifically, they stalled the spread of European livestock. Instead of raising flocks and herds, they let the missions and Mexicans do it, and then raided them. It was only after the Apache wars sequestered the remaining tribes onto reservations in the late 19th century that livestock exploded and the regional fire load crashed. Across the Southwest the slow-combustion of cattle in particular beat back the fast-combustion of open burning. The break in fire history is as abrupt and distinctive as the abandoned dwellings of the Mogollon and Anasazi. Like the Apache, free-ranging fire retreated to reservations or isolated locales spared the general crush of cattle.[4]

San Carlos was first gazetted as a reservation in 1872 and became a collective for various Apache bands that had little in common save their language. The Yavapai, the Tontos, the Chiricahuas, and fragments of other bands of the western Apache were rounded up like mavericks and put into a common corral. The experience was wrenching. Most had no ties to their new estate. They had no cultural continuities that could bond them to the land. The old ways often had little value. They would have to rebuild a new culture, find new stories to pass on their inherited wisdom, relocate from their sacred mountains to the desert hills of San Carlos's lowland administrative post. The unsettled demography and scrambled culture meant that whatever fire practices they had known in the past would have to be reconstructed in a setting over which they did not control the basics of their existence.[5]

The gathered groups were fed ration beef, and then granted cattle to slaughter or raise. It made sense to grow the beef on site, so herds came to San Carlos itself, as throughout the Southwest, although most arrived through trespass from herders outside the border. Some order was instilled by establishing a leasing system for the Anglo ranchers, which continued until 1933 when the tribe began reclaiming control. San Carlos's flatlands were prime grazing land, and the herds were large; the resulting ranches were among the last driftwood deposited from the storm surge of the Texas cattle industry; and here, like nearly everywhere, abusive overgrazing was the norm. A ruinous cattle rush that had degraded landscape after landscape in the American West came to the desert grasslands, forest savannas, and high prairies of San Carlos. The fire regime that had prevailed for centuries, however modified by the Apache, was eaten and trampled away.[6]

At the same time notions that fire ought to be actively suppressed added to the pyric loss. The Bureau of Indian Affairs developed forestry programs to promote timber industries, and foresters at San Carlos did what foresters everywhere did: they fought fire. Between intensive grazing and active firefighting, the lavish ignitions could no longer roam with the insouciance of black bears or the free flow of the wind. Increasingly, the land became, for fire, a *tierra desplobada*.

The BIA established a branch of forestry in 1910, but it was not until the reforms of the Indian Reorganization Act of 1934 that San Carlos felt its presence. The CCC added muscle, committed to fire control as much as to erosion control, building fire roads and towers as they did rock check dams. The tribe began reclaiming ranching leases and looked to forestry to enhance its economic development. In the 1950s commercial logging developed. The postwar era set San Carlos on the path that led, ultimately, to the April 23 briefing.[7]

Its foresters were conscious that they managed tribal lands "in trust" and accordingly transplanted the prevailing practices of their day to San Carlos. But even as modern forestry arrived, its norms were being challenged. When he came to Arizona in 1948 as regional forester for the BIA, Harold Weaver gave voice to the concern that fire exclusion was disrupting the land as fully as the overgrazing that accompanied it. He also discovered that parts of San Carlos had apparently been spared.

He found in Malay Gap, along the Mogollon Rim, at the far northeastern corner of San Carlos, a place that seemed relatively unscathed by hoof, ax, or removed flame. There, fire-scarred ponderosa suggested that surface fires had returned at least every seven years over several centuries and had bequeathed a forest so magnificent that "it is hard to see how she can be much improved on." The fire challenge at San Carlos was to propagate that old regimen of burning, not to find better ways to knock it out. Weaver urged controlled burning of the sort he had helped pioneer in the Colville and Warm Springs reservations of Oregon and which Harry Kallander, north of the Black River at Fort Apache Reservation, was introducing. But just as simply removing cattle dramatically unburdened the land, so ceasing to suppress the fire that nature so lavishly strewed about the landscape, would help. In fact, lack of resources meant that places like Malay Gap could perpetuate something like the old ways. Major burns had washed over Malay in 1943 and 1946, just prior to Weaver's tour.[8]

Mostly, though, prescribed fire meant burning the slash generated by the emerging timber program. Instead of adopting the Weaver agenda, San Carlos committed to fire control, which paradoxically acted not as a drain on the tribal economy but as a stimulant. The San Carlos Apaches heavily staffed the Southwest Forest Fire Fighters program, fielding as many as 800 men a season. Those paying jobs were a significant source of otherwise scarce income; they did for the feeble money economy of the reservation what migrant farm labor and remittances did elsewhere. At the same time, San Carlos forestry built up its own suppression infrastructure. It acquired engines. It erected lookouts and founded an aerial reconnaissance operation. It built a modern radio network. It developed a spider web of fire roads. It had access to bulldozers and air tankers. It built a fire camp at Point of Pines, complete with a helitack crew. In 1991 it founded the Geronimo Hotshots, over which the tribe assumed control in 1996. Prescribed burning was mostly limited to activity fuels, which is to say, logging slash.

Even as reforms swirled in the convective plume that promised a national fire revolution to reinstate fire, a trend with which its natural and ethnographic history would seem to align perfectly, and for which Harold Weaver had become a prophet, San Carlos hardened its commitment to serious fire suppression. The 10,000-acre Black River wildfire of 1971, not a scheduled prescribed burn on Ash Flats, defined the thrust of the program. In 1972 Tall Timbers Research Station sponsored a field tour to San Carlos and Fort Apache reservations, with Weaver among its company, and noted that any attempt at prescribed burning was failing to keep pace with the wildfire threat; the failure to install a vigorous fire restoration program would only lead to more blowouts. Yet it was a hard choice. Fire suppression was the national norm, and if nothing else, firefighting meant money. Its crews were one of the few exports the tribe had. Big burns brought in big bucks, and while prescribed burning and managing wildland fires would still hire crews, one of their objectives was to reduce dollars spent.

So San Carlos missed the national fire revolution. Not until the revolution revived in the mid-1990s was there pushback against simple suppression, and that proved tricky because the fire program needed all the resources of the suppression program while redirecting them toward a more holistic strategy. An understory burn around Point of Pines and rangeland burning (to dampen encroaching cholla) put controlled fire on the ground for the first time since Weaver. The prime mover, Bob Gray, recalled that the "beating I took from my own administration, the BIA, the wildlife biologists, etc. was brutal—all wanted the experiment to fail; my only allies were the elders who understood the importance of fire." But he had the elders, and through them, the tribal council, and the example (and advice) of Weaver. The Laboratory of Tree-Ring Research began fire histories. It helped that the suppression program, though large, had become moribund, and in 1995 the tribe took control of the Geronimo Hotshots from the BIA, and then upgraded suppression standards and rewrote the fire plan to accommodate some cross-border burning with the Apache-Sitgreaves National Forest and to establish a large natural fire zone in the eastern lowlands below the Nantac Rim. Some 35,000 acres across the constellation of San Carlos biotas burned.[9]

The 2000 National Fire Plan funnelled more funds toward equipment and fuels.[10] But it also required a plan. The plan would have to reconcile fire practices with land use, which is to say, the forest management plan (and environmental assessment, and Tribal Strategic Plan). The new task was to transform policies that made possible an active program of fire restoration into one that effectively mandated it. After 2001, however, the latent prospects of policy met the right personalities and right politics, and the resulting wildland fire management plan of January 2003 announced a new era.[11]

The fire program picked up the pace of slash burns and the large experiments of the previous decade, but quickly appreciated that more was needed. If fire spread is a question of surfaces, fire management is an issue of edges. The more surface relative to volume a fuel particle has, the faster heat and moisture transfer, such that small fuels burn more readily than large ones. Similarly the more edge relative to area a burn plot has, the more complex its control. The fire program burned 600 acres at Baskin Tank in 2006, and 2,500 acres at Dove Tank; but even amid logging slash or chained juniper, approval could take several years and cost serious dollars. And they were only treating new fuels they created, not legacy landscapes. The San Carlos fire program had to enlarge the size of the sites beyond individual plots and quicken the tempo of treatments.

Most clustered along the Nantac Rim. Beginning with reconnaissances in 2005, and plans in 2007, they turned to the Hilltop region. Between 2008 and 2010, with help from aerial ignition, they burned off three big blocks that added up to 14,000 acres. In 2009 they confined and contained the 20,000 acre Bear Canyon fire, even as they attacked 45 new starts. In 2011 the Maggie fire added another 5,000 acres. In 2012 the Trail and Shorten fires brought in 8,000 acres more. Managing fires in this way reduced costs by an order of magnitude. Meanwhile, like the rest of the Southwest, bleached by a long-wave drought, wildfires blotched woodlands north and south of the rim on the order of 500 to 4,000 acres.

Then came 2013. Early on they set three prescribed burns, a few thousand acres each in woodland savanna, along with the complex 13,000-acre

Pine Salt burn around Point of Pines. The wildfires arrived on schedule. An "annoying" fire burned through salt cedar near Bylas along the Gila River. Two fires broke out in the high country. The Fourmile fire burned toward the reservation's eastern border through a "scabby transition zone," not easily attacked nor worth risky suppression. Fire managers backed off to roads or barriers, burned out, watched, and otherwise confined and contained. The Creek fire on the north flank of the Nantac threatened the Dry Lake Lookout and a remote automated weather station, some commercial timber, and Point of Pines; but even in the height of fire season, there were options other than going toe-to-toe, and fire operations backed off, burned out selectively, used previous burns as cold trails, called for some air strikes near developments and canyons where the fire might bolt, and generally herded it to good effect, though its fire behavior specialist, Bil Grauel, then handling the wildland fire decision support system for the burn, complained to Duane Chapman, fire management officer and former superintendent of the Geronimo Hotshots, that he "couldn't model the damn thing if he [Duane] was going to herd it all over the landscape." The Creek fire blackened 18,000 acres. That year San Carlos had the largest fires in the state. What Grauel said about the Fourmile fire epitomized the season: "Along with the fire, they decided that 14,000 acres was about right." That statement's odd phrasing, and the attitude it conveyed, could stand for the program.[12]

Fire management at San Carlos is under the Mogollon Rim and under the national radar. Like all programs it has its liabilities and its assets. How they balance says a lot about how the program actually looks on the ground.

The liabilities are numerous and obvious. Many, like patchy support, are shared by all fire organizations, and like metastasizing juniper woodlands, by those throughout the western United States. Some, like shortchanged funding relative to federal neighbors, belong to reservations and the oft-poisonous codependency between tribes and the Bureau of Indian Affairs. A few are specific to San Carlos. The tribe's instincts to turn inward into a kind of cultural as well as economic autarky. The long reliance on fire suppression as a seasonal revenue stream. The lack of

Apache traditions for landscape-scale burning. The attitudes of neighboring landowners more cautious about free-ranging fire and far-reaching smoke. The collapse of the Southwest Forest Fire Fighters program as trainees failed physical and drug tests. The way social pathologies seem to channel into land degradation.

But the program's assets may be greater. Over the past 20 years, livestock numbers have shrunk dramatically, and today are perhaps 20 percent of what they were previously. That has freed up grass to carry surface fire. Tribal autonomy allows it a freedom to maneuver not available to the national forests to its east, west, and south. The lack of major cities, or even exurbs, means the interface is a minor concern. If Apache culture lacks traditions of burning, it also has a tolerance for allowing natural processes to work their own destinies, which can translate into flexibility in handling the fires nature starts. A small timber program means the tribal economy can absorb some burned stands as a price of building resilience into the land and preventing savage wildfires; reduced cattle stocking means it can accept burns that take away winter range. The land is so fire-sated naturally that reform does not depend on fire lighting; it's enough to modify fire fighting. Even isolation and lack of attention has its merits. If insularity means San Carlos doesn't get much outside support, it also means it doesn't get much outside scrutiny.

It means San Carlos lives for the present in a sweet spot. It can do things its neighbors can't. It can turn its relative poverty into a wealth of opportunity. It can replace ever-more-encumbered prescribed burning with hybrid fires in which burnouts and free-burns from natural ignitions fuse. In the past, decisions about fires were boxed in by the demand to suppress them as quickly as possible. Now starts like the Skunk and Basin fires allow fire managers to draw their own boxes. That grace period won't last, but it grants San Carlos a time to stabilize its program so that, when something does go wrong, as it inevitably will, its fire program will survive.

On June 6, 1946, Emil Haury of the University of Arizona, one of the premier archaeologists of his day, was busy erecting a field camp at Point of Pines, what would become over the next 14 years a celebrated summer training site for students of southwestern archaeology. At noon he received a call from Paul Buss, the San Carlos forester, with a request to aid in fighting a fire at Malay Gap. "An unwritten law in the wilderness country," Haury sternly noted, "says that when the forest begins to burn, all able-bodied men must pitch in to help bring the fire under control, no matter what they are engaged in at the moment." Six members of the company made the trek to Malay, "about as remote as any place could be."[13]

Amid a "totally strange country, extremely rough, and heavily timbered," Haury and his students struggled to cut line during the late afternoon. One student found himself trapped in a "fire circle" and barely escaped with his life. That night, with aching muscles, they struggled to sleep while immersed in smoke "from which there was no escape." The "eeriness and gloominess of the situation, the crashing of burning trees, the sharp riflelike cracks of exploding rocks around us, and, the next day, faced by a roaring wall of fire racing up the hill to our temporary camp, were enough to sear the words *forest fire* deeply in any mind." Meanwhile, several score Apaches on the scene exhibited "no special concern." Without strong direction, the archaeologists "lost heart" and asked to be released. From Point of Pines they could watch the smoke billow up for some days, apprehensive and grateful that it was so far away. "I will say," Haury concluded, "that no baptism by fire, literally, was ever more exhausting and frightening."[14]

What happened that inaugural summer was repeated in the years that followed more often than Haury wished. He came to view fire as a serious inconvenience. In 1950, the group was called out "frequently" to lightning fires, which "sorely interrupted" their work schedule. In truth, fire became the secret catalyst for some of their major discoveries. They tracked the history of the built landscape through hearths. They traced the historic mingling of peoples through cremations. In 1950, that summer of fire, they followed gopher mounds to charcoal and carbonized corn, which led them to "the very thing we were looking for," evidence of a "conflagration." Charcoal is nature's great preservative. When they excavated a pueblo with 20-plus rooms burned, they had a mother lode of how the Mogollon culture lived, what they ate, and when they flourished, and in the form of scorched skeletons who they were. The next summer they attended fire school, had even more callouts for fires, and at one point had to devise evacuation plans. Fire good, fire bad—student archaeologists were learning the real lesson of Point of Pines.[15]

They uncovered some startling finds. In the 13th century, the Mogollon and Anasazi cultures had met in an awkward mingling and occasional fusion best exemplified by the emergence of a square kiva. The kiva was Anasazi; the squared corners, Mogollon. The square kiva came to stand for the prehistoric world of Point of Pines, and it might foreshadow the modern world of fire management there as well. Kivas come in circles; fire, in triangles. The Mogollon built out of rock, the moderns out of the triangle of ingredients that make fire—heat, oxygen, fuel; terrain, vegetation, weather; fire lit, fire fought, fire herded. Prehistoric San Carlos managed to square the circle of their times. Contemporary San Carlos is squaring the triangle of theirs.

As the Mogollons rendered an idea into stone, so San Carlos fire must transfigure its innovations into institutions. It has happened before. Twice, in fact, San Carlos has succeeded in transforming a mangled scene into a model system.

The first involved the wreckage of its heritage of free-range ranching. In the 1930s, as the tribe began reclaiming leases, it also became the site for the first national experiments in artificial insemination under John Lasley of the University of Missouri. The numbers of scruffy stock came down, the quality of the remaining Herefords went up. By the 1950s a model ranch, the R100, combined breed improvement with range management on Ash Flat under the direction of the University of Arizona. Meanwhile, Emil Haury helped turn generations of free-range pot-hunting into an academic discipline, and underwrote an archaeology suitable for a research university, again the UofA. Its peculiar isolation made San Carlos reservation ideal for both purposes.[16]

Now it may be fire's turn. Abusive practices—in this instance, fire exclusion—are being transmuted into an exemplary exercise in fire restoration. The old Point of Pines fire camp, once a high scorch mark in the era of firefighting, could be remade into a training ground for fire lighting. The Prescribed Fire Training Center in Florida has as its motto Every Day Is a Burn Day. Active prescribed burning is far trickier in the West, but San Carlos doesn't have to wait for the cumbersome protocols of fire by prescription. The land is sated with natural ignitions that only need to be brought under a system of management. San Carlos doesn't have to rely on fire lighting to reinstate a more resilient fire regime. It need only modify its acquired habits of fire fighting.

And that is what the fire staff at San Carlos did with Basin and Skunk. They focused first on the more interesting Basin fire. They drew the box, and the contours of their burnout looked for a while like a folding protein as it absorbed old prescribed fires and patches of wildfire before stopping at 6,018 acres. When weather moderated about a week after Easter, the staff ignited the 1,228-acre Point of Pines prescribed burn to clean out thinning around the camp before rotating those crews into the 524-acre Bee Flat burn to restore juniper savanna. A wildfire started at Willow and was suppressed at 21 acres. Another started in grassy scrub at Rimrock and was boxed in at 1,826 acres. The Barlow fire broke out along the main road through the Nantac, burned hot, and was fought hard to a tough 1,483 acres. And the Skunk began to move.[17]

It was 30 acres at the end of April, and 80 on May 11. Then a mild cold front passed over and strong northers pushed the fire to 1,822 acres and put flame on the grassy plains to the south of the Nantac Rim. Here it found a fuse it could burn along while prevailing southwesterlies could drive it onto the rim. The fire was filling in a long blank spot in the need-to-burn map of San Carlos. From the 11th to the 21st the Skunk fire rose and fell, making daily runs as high as 6,336, 9,248, and 4,254 acres. The San Carlos staff ran along beside it, using roads and burnouts like drovers holding a stampede. They put a crew in to prep a repeater station before the flames arrived. The fire front continued to spread, unusually for the Nantac, to the northwest. San Carlos called for a Type III short team to help with logistics, although the newcomers had to be constantly educated into what San Carlos wanted (and didn't want, an expensive air show). They thought Kidde Creek might hold the progression; it didn't. They continued to burn out along 1500, the main road along the rim. They downsized to a Type IV crew, used some helidrops to help hold the burnouts and called in some retardant from small air tankers. On the 27th the fire rushed over 4,879 acres before pausing. At Rocky Gulch they pinched off the front and began burning out the interior with aerial ignition. That accounted for the last big burn—14,087 acres on June 2. The fire perimeter now ranged over 92 miles, roughly 22 miles long and 8 miles wide. When the smoke settled, the Skunk fire had blackened 73,622 acres and tied in with the 2009 Bear Canyon and 2004 Upshaw fires to the northwest and the 2013 Creek fire to the southeast. The summer fire season—with its normal dry-lightning fire busts—was still weeks away. Staff were already wondering how to handle the reburns that would be essential to turn a spate of fires into a functioning fire regime.[18]

Meanwhile, lightning had kindled the Black River Tank fire along the border with Fort Apache reservation. Fort Apache tried to emulate the San Carlos strategy, but instructions were confused, and an air attack operation saturation bombed to hold the burn to an expensive 3,244 acres. What San Carlos had accomplished depended on the dynamics of its people, not just the dynamics of fire burning through pine, juniper, and grass.

Fire behavior is fire behavior and universal, but behavior toward fire is specific to cultures and not transferable with algorithms. Box and burn is not a simple tool, like a Neptune air tanker or a D6 caterpillar that can be dropped into any landscape. It is a negotiation between fire and fire managers. Like all things human it has to be learned, but unlike many it is not something easily taught.

A REFUSAL TO MOURN THE DEATH, BY FIRE, OF A CREW IN YARNELL

I shall not murder
The mankind of her going with a grave truth . . .

—DYLAN THOMAS, "THE REFUSAL TO MOURN THE DEATH, BY FIRE, OF A CHILD IN LONDON"

ON THE 30TH OF JUNE 2013 a fire blew over the Granite Mountain Hotshots outside Yarnell, Arizona and left 19 dead. Three months later, on September 30, a formal investigation released its findings. The inquiry focused on the mechanics of fire behavior and how the Granite Mountain Interagency Hotshot Crew might have understood their "situational awareness," which is to say, how the crew recoded the information they were given with what they saw for themselves. Instead of ascribing blame, the investigative team sought to appreciate how the hotshots engaged in sensemaking in an effort to explain decisions that, to nearly all observers, made little sense. But the need for sensemaking extends also to the meaning of the fire for American culture at large.

For anyone conditioned to read landscape for fire behavior, Yarnell Hill is a Google of clues ready to be coded into the existing algorithms of fire behavior. The fundamentals point to fuels of mixed brush and grasses, parched by seasonal drought, to the terrain of Yarnell Hill, and to record temperatures, blustery winds, and the downdrafts ("outflow boundary") from passing thunderheads. There is nothing in the reconstruction of the fire's behavior that suggests it was anything other than a high-end variant of what happens almost annually.

What made a difference was that the collective will of a hotshot crew crossed that flaming front. The reaction intensity that matters is fire's interaction with American society. The behavior we want to explain is the crew's, and what their death signifies, and for that we must look outside the usual fire-behavior triangle and into that triangle of meaning framed by literature, philosophy, and religion. The fire came as a tear in the space-time continuum, opening into a void for meaning. There are fires that belong with science, fires that stride with history and politics, and fires that speak in the tongues of literature. Yarnell Hill is a literary fire. It's a fire for poets and novelists, and maybe the stray writer-philosopher.

In a profile of James Pike, Joan Didion observed that he might be understood as a "great literary character" like Jay Gatsby. So, after the reports are filed, the lawsuits settled, and the scientific interrogations published, we may well linger over the Yarnell Hill fire as a great literary moment, for which character, conflict, and plot serve as the fuel, terrain, and weather resolved into the blowup of tragedy.

The greatness of the fire does not lie in its physical behavior or scale. Hannah Arendt famously spoke of the "banality of evil." It is likely that the mechanics of the 8,000-acre Yarnell Hill burn will prove equally banal, not with active evil but with an unsettling emptiness. It is not what happened but what it means that mesmerizes the public imagination. Giving story to that sentiment will be the task of literature.

Nor blaspheme down the stations of the breath
With any further
Elegy of innocence and youth.

When it studies fire behavior, the fire community reaches for models; so, as it ponders the fire's meaning, it will need to search for narrative templates. The vital parameters are a wrathful gust of wind and a devouring fire, the immolation of 19 men, a lone survivor, a landscape of ambiguous purpose as a determined brotherhood chases flames by a nearly deserted town. These facts the community will try to reconcile with fire behavior and to the interpretive models that it uses to account for why fire suppression today is dangerous. The problem is that its prevailing models don't fit.

Expect that the community will turn first to Norman Maclean's Talmudic *Young Men and Fire*. Attention will focus, in particular, at the point where science met literature, the study of comparative behaviors, the fire's and the crew's, that Maclean's inquiry inspired from Richard Rothermel. At Mann Gulch the steep terrain quickened the fire even as it slowed the smokejumpers. Where the curves of their differing rates of spread crossed, the crewmen fell. In Maclean's hands the mathematics became the narrative lines of a Greek tragedy in which fire and crew each did what they were destined to do. Bob Sallee and Walter Rumsey were high enough on the slope that they could just evade the flames. For the others, only Wag Dodge's escape fire—like a deus ex machina—interrupted the logic of fate.

Yarnell Hill will become another "race that couldn't be won." But the analogy stumbles. The smokejumpers at Mann Gulch had no choice: they were trapped in a closed basin, almost a chute, and would perish unless they could outrace or outsmart the flames. The hotshots at Yarnell Hill were safely on a ridge, in the black, and chose to race with the fire by plunging downhill into a box canyon thick with boulders and brush. Theirs was an act of volition denied the jumpers at Mann Gulch. At Mann Gulch the fatal numbers were coded into the scene at its origin. At Yarnell Hill the Granite Mountain Hotshots did the calculations and added the sums incorrectly.

The numbers tempt: they are hard facts, recorded in the landscape, not unknowables embedded in the nebulous "sensemaking" of mind and heart. So one narrative will turn to explanations for the fire's "extreme" behavior from outflow winds, box canyons like thick chimneys, and the boiling dynamics of plumes. The story will look, in particular, to long-unburnt fuels and especially to extended seasons, record temperatures, and climate change. The explosive Yarnell Hill fire will become another signature of the Anthropocene's new normal. The loss of an elite crew will be tallied as part of the cost of ignoring global warming.

It was hot, dry, and windy, but it's always like that in the early summer lead-in to the monsoon. Central Arizona has known higher temperatures, stronger winds, and deeper droughts. When thunderheads collapse, particularly in the season's first storms, those blasting downdrafts drive flames as they do city-enveloping haboobs. A similar outburst drove the 1990 Dude Creek fire (eastward at Payson, a sister city to Prescott) through a crew and killed six. The flames at Yarnell Hill leaped and spotted through grass and brush—combustibles ideally suited to react quickly to wind. Conditions on June 30 were not beyond the region's environmental or evolutionary scale. We don't need climate change to account for the fire's behavior.

Already, the community is turning to that other great narrative template invoked to describe the contemporary scene, the geeky-named wildland-urban interface. The Granite Mountain Hotshots were putatively on the scene to defend Yarnell, an exurban enclave that may have been indefensible and in any event was mostly evacuated. It's unfair to demand that fire crews risk their lives for property. That burden belongs to the community. If they build houses where fires are, they have to live with fire.

But Yarnell frustrates this generic model. The town sprang up during an 1860s gold rush. It was platted 50 years before Arizona became a state. It survived by being repopulated, most recently by retirees. Whatever the firescape at the time of founding, it was undoubtedly scalped by the miners, who burned off the cover to expose outcrops, cleared any trees and shrubs for firewood, and brought in meat cattle that stripped away the grass. The existing scene is the jumble of recovered pieces. The town was not plucked down amid combustibles; the firescape grew up around the town. In fact, by the early 1950s the state of Arizona was actively promoting Yarnell as a retirement community, which is what it became.

The dispersed outliers were mostly indefensible, so there would have been little justification in trying to shield them, particularly with fire bearing down imminently. The Granite Mountain IHC was outfitted with hand tools for cutting fireline, not with hoses, pumps, and shielding for defending structures. Before they left the ridge, they were working a free-burning fire; they were assigned to establish an anchor, not to protect structures. This was not an Esperanza fire in which flames washed over an engine crew positioned to defend a house. The Granite Mountain IHC did not perish, arms locked, standing between the flames and homes. They died in a box canyon into which they had voluntarily hiked.

The other template—everything has to come in threes—is that legacy agencies are unable to overcome their culture of suppression. They fight fires where and when they shouldn't because they know no other way to respond; the endless roster of shelter deployments, near misses, and fatality fires is the inevitable outcome. Even the shock of the 1994 South Canyon deaths, the prospects for civil and criminal penalties, and the emphatic edicts from above have failed to dislodge that culture. It invites risk-taking that leads remorselessly to Thirtymile Canyon, Cramer, and Yarnell Hill. The rules keep being broken—must be broken to satisfy the nature of the beast. Implicit, too, is a tint of gender bias; the Granite Mountain crewmen were all young men engaged in what could seem an extreme sport.

Yet again this particular fire turns such understandings into shades of gray. Granite Mountain IHC was not a Forest Service or BLM crew: it was proudly, defiantly, the only nonfederal IHC on the national register. Undoubtedly it absorbed elements of traditional mores and camaraderie, but perhaps without the institutional checks that have been cultivated over the generations since 1994. It belonged to a city fire department. It absorbed at least as much culture from urban fire-service expectations as from wildland agencies. It was fighting a wildfire on Arizona state lands, under the auspices of the Arizona Department of Forestry, the youngest state forestry bureau in the United States, established the last year fires burned in Peeples Valley. There were other crews on Yarnell Hill, including federal hotshots. None of them put themselves into the path of the fire.

In the least valley of sackcloth to mourn
The majesty and burning of the child's death.

If the traditional templates don't apply, what might? If this is a literary fire, they will come from literature, and in truth possible models seem to leap from the scene. There are those that help arrange the particulars into a story, and those that invest that narrative with the kind of meaning that speaks to the extraordinary reach of interest that this fire has sparked. They move the tale from physics to metaphysics.

Look, first, to *Moby-Dick*. The pivotal—the Ahab—character is surely Eric Marsh, cofounder and crew superintendent, not because he is mad or malevolent, but because he is driven, and his role, and whatever future inquiries determine to be the chain of events, this is his *literary* role.[1] Wildland fire was his obsession: he helped transform a brush-clearing crew into an IHC, he established the Arizona Wildfire Academy in his living room, he modeled the crew after his own character. They were all young, male, and Anglo. They had something to prove: they were a proud "oddity," he had written some months earlier, a city-sponsored IHC among a federal-dominated workforce; a "mystery," to city coworkers;

"crazy," to family and friends. They prided themselves in showing up "to a chaotic and challenging event, and immediately breaking it down into manageable objectives and presenting a solution." They did not just call themselves hotshots, they were "hotshots in everything that we do." They "loved" the life they had chosen. They "managed to do the impossible."[2]

On the afternoon of June 30 Marsh was a division superintendent, but he left the line to scout and then moved the crew away from a blackened ridge and back into the action. He showed an élan and initiative that in many circumstances of life we would applaud but what here looks like an obsession. They were safe. He took them to the flames. His drive became a fatal flaw and carried the others with him. The crew, even the Starbucks among them, follow, all caught up in the chase. So at last that great white whale of a fire turned on them, and left one survivor, a solitary witness to proclaim, after the Book of Job, "And I only am escaped alone to tell thee."

Still, tragedies abound. What made this of interest in Toronto and London as well as Phoenix is that the fire dramatized with sudden and graphic violence the question of what purpose if any informs our existence, whether our lives reflect the workings of a plan or of accident. Look, in this case, to Thornton Wilder's *The Bridge of San Luis Rey* to account for the awful coincidences and the quest for patterns in the void. "On Friday noon, June the twentieth, 1714, the finest bridge in all Peru broke and precipitated five travelers into the gulf below. . . . People wandered about in a trance-like state, muttering; they had the hallucination of seeing themselves falling into the gulf below."

Why this crew? at this time? in this way? Why, as residents of Yarnell asked, were some houses spared and others burned? Was there a hidden order, which is to say, a deeper providence in the tragedy, or was it just an arbitrary collision of actions with no more design than the scatter of summer cumulus? The Granite Mountain IHC had for its original logo a pair of flaming dice that always came up seven. This time the dice rolled snake eyes.

What the *Bridge of San Luis Rey* model also suggests is that the way to narrate the meaning is not directly under the gaze of an omniscient narrator arraying events within GIS grids and plotted along timelines, but through the quest for their significance. Norman Maclean made that pursuit personal: *Young Men and Fire* became the story for his own search for the story. Thornton Wilder refracted that inquiry through Brother

Juniper, whose pursuit of and meditation on the gathered facts leads ultimately to an auto-da-fé, the burning of self and book at the stake.

Beyond metaphysics and theology lies the tangible grief of the survivors. They want significance ascribed to the sacrifice. They want their loved ones honored and valorized. This, too, is an old provenance of literature, and it offers a full gamut of consolations, from doggerel and sermons to narrated emotion fused with artful intellect. For this task ponder James Agee's *Let Us Now Praise Famous Men* to appreciate how to ennoble ordinary lives without sentimentalizing their fate. And at its most challenging, read Dylan Thomas's inextinguishable "A Refusal to Mourn the Death, by Fire, of a Child in London" for the cold, obdurate turning inside out of emotional trauma in the face of a tragedy of innocents.

For all its splendor and pathos, American fire has no novel to match its stature in the American scene. It has some powerful nonfiction, most notably Maclean's *Young Men and Fire*, which has many imitators but stands alone, itself a survivor and posthumously published witness. But the fusion of fire and art remains unmet. In Yarnell Hill, however, the American fire community has the themes and latent structures for a great work of literature.

That doesn't guarantee it will happen. Five books are promised on Yarnell Hill (two published, to date). My guess is that the enduring voice of the tragedy, however, will come long after the event, much as the great novel of the American Civil War, *The Red Badge of Courage*, was published 30 years after Appomattox by someone who was six years away from being born when the war ended. The author will be a writer who recognizes that this is not just human-interest journalism, a gripping story of a disaster, a procedural on fireline behavior, or a political parable about misplaced national priorities, but someone who appreciates the fire as a great literary character whose meaning must depend on the ambiguities of art to extract significance from the indecipherable. Not just someone who can see patterns in the flames and hear cries among the roar, but someone who can say with Dylan Thomas that

After the first death, there is no other.

EPILOGUE

The Southwest Between Two Fires

FIRE IN THE SOUTHWEST is a rectangle.

Fire behavior and fire history depend on terrain, climate, biotas, and people, and the people have organized themselves into four cultural hearths that border and penetrate the region. Each hearth is a nation or a shadow state-nation that once was independent but retains its own origin story, political culture, economic realm, and identity. To the east lies Texas; to the west, California; to the north, Utah, the rump of old Deseret; and to the south, Mexico. All converge, like the Rockies, the Colorado Plateau, the Basin Range, and High Plains, on the region, and manage to both coexist and not merge into a common meld. The Southwest has no single informing identity: its character derives from the mutual jostling of the others. It holds all the fire themes that define the national scene, but what defines the region is how those themes mutually reside relative to one another—sharp, not blurred; separate, not fused; visible, not hidden or disguised; preserved, not erased.

What makes the Southwest distinctive is the sharpness and clarity of its contrasts. Gorge drops instantly from mesa; metropolis and cliff dwelling reside almost within arm's reach; thunderheads billow up singly against otherwise empty blue sky; industrial combustion abuts free-burning flame. Wilderness burning with the oldest of fires next to conurbations run by the newest technologies of combustion. In the dry air and exposed rock nothing seems hidden, yet the land is full of surprises.

Northern New Mexico has an aura of magical realism: every turn throws up a new vista. The most celebrated of Arizona's landforms, the Grand Canyon, is virtually invisible until you stand on its rim.

In its human geography, as with its natural, the borders are delineated and the contrasts sharp. The time-honored solution to coexistence is to allot each feature, group, or story its own space. Instead of assimilating the lot into a collective melt, the pieces find ways to cross borders without dissolving either source or sink. Even the elderly have their enclaves like Sun City, porous for commerce and politics, but impermeable to residence. The strategy has its logic: in fact, some observers see the entire United States sorting itself out along similar lines of class, ethnicity, and race. The Southwest just shows the trend with unfiltered transparency.

In cultural matters, as in fire, abrasion can be a point of ignition. The fires—and even more problematically, their smokes—must stay within their own reservations and castes. The contested areas are those where groups and enclaves overlap, or one spills over into others; where wild and working landscapes compete for common space; where landscapes organized around free-burning flame meet those informed by internal combustion.

As with peoples, so with lands, and as with lands, so with fires.

The region's arts and sciences have seized on its elemental matrix. Its rocks have contributed to world geology and its minerals to the global economy. Its waters, both surface and subsurface, have established paradigms for national politics in natural resource management. Its atmosphere has defined much that is distinctive about regional science and art, from its bevy of telescopes to the founding of a tree-ring lab, whose initial task was to identify climatic cycles tied to sunspots, and from the startling clarity of William Henry Holmes's drawings of the canyon country to the unblemished starkness of the Taos School. And its fires—its fires continue to shape national thinking.

The region's literature, whether hack or high, builds on those elemental contrasts, particularly the rubbing of one group against another. Zane Grey's *Riders of the Purple Sage* hinges on the contrast between Mormon and Gentile, and ends with hero and heroine escaping to a hidden enclave that places them beyond further conflict. Oliver Lefarge's *Laughing Boy* explores the lethal clash between Navajo and Anglo. Willa Cather's *Death Comes for the Archbishop* traces the awkward reconciliation of high culture with the folkways of Pueblos and old-resident Hispanics, not by absorption so much as mutual acceptance. Aldous Huxley updated the clash between Taos pueblo and the modernity of Mabel Dodge in *Brave New World.* Tony Hillerman wrote an oeuvre in which different peoples can see and interpret the same events through separate cultural prisms with equal acuity; a murderer who changes his identity, for example, does fit the Navajo definition of a witch. Each group adjusts to the other without surrendering its own rendering of the world. The best of these accounts have shaped the national literary canon.

In contemporary times the aperture of literary imagination has widened to include environmental themes. Landscape has always loomed large in Southwestern art, not just framing action but threatening to overwhelm it; now it is becoming itself the subject. (Think Ed Abbey and the Tucson school.) Here, too, conflict hinges on a contrast between irreconcilables—in this case between a land usage committed to funneling goods into the national economy and a land ethic based on nonanthropocentric values. Such notions have created new categories of land use, among them legal wilderness and reserves for endangered species. The Gila National Forest housed the first primitive area. The Center for Biological Diversity, based in Tucson, pioneered aggressive use of the Endangered Species Act.

On the surface there is little chance to reconcile political philosophies founded on such fundamental differences. There is no middle ground because the one demands a human presence and the other denies it. Together with the exurb, the legal wild has intruded into the Southwestern matrix with the power of the Athabascan migrations, the American invasion, or the contemporary surge of border crossing from Mexico. It has compelled a reconstitution of extant enclaves. Amid such abrasions and uncertainties fire flourishes. It all makes good literature but awkward politics and even murkier fire management.

Fire can be hard to keep within bounds. You can't dig it away or shove it into mounds like dirt. You can't channel it or impound it like water. You can't hold it in reserves by court order and gunpoint. Instead it tends to ride with the winds. If it is straying from its politically appointed reservations, it is only obeying its internal logic. Unless there is an abrupt change in fuels or terrain, the flames won't distinguish between wilderness and exurb. Some borders make sense, some don't, but the issue before fire management is to reconstitute those bounds into manageable edges.

The Southwest suite remains a mosaic, but one of history as much as geography. Some of the tiles of the old mosaic endure, some have broken and been replaced. In recent times two phase changes dominate. The first happened when the indigenes (and the indigenous fire regimes) were broken, suppressed, and scattered into reserves. The second arrived with the post–World War II boom. The Old Southwest had stood shoulder to shoulder with open fire, until its herders trampled the flames away and its loggers hacked off the woods and fire setting became subject to criminal prosecution. The New Southwest was predicated on industrial combustion to power its cars, air conditioners, and electronic gadgets. It has revived open flame, although as much by accident as by choice, and by replacing logging camps and ranches with exurbs and campgrounds it has complicated fire's managed reintroduction. Instead, the Southwest suite is one of barely controlled industrial combustion and largely uncontrolled feral fire.

Today, the Southwest's fires are no longer confined within genres or enclaves or the approved master plans of parks and forests. The distinctive pieces endure. Fire just sweeps over them all.

Quietly, without the fanfare that accompanies innovations from other regions, the Southwest is exploring solutions.

They are easiest where the land belongs to one owner and is self-contained—the sky islands are good examples. Problems arise where outside interests pluck telescopes or recreational homes or research centers onto and among the wilds, or where the borders delineate keenly different land usage. The FireScape Project seeks to create a fire commons among those owners, establishing a single landscape for fire's management. In contrast, the Four Forests Restoration Initiative spans four national forests which gives it a single owner, though four administrators; and it applies to a forest as homogeneous as any in the country. Still, there are borders to defend, some pockets of settlement and summer homes nestled like measles spots, and quarrels over how much of the land should be working and how much wild.

The most daring of the experiments may be the Malpai Borderlands Group. Here borders are everywhere—and everything. The lines are drawn between Arizona and New Mexico; the United States and Mexico; the Forest Service, the BLM, the Fish and Wildlife Service, and the State of Arizona; and private landowners, from the Nature Conservancy to ranchers, some holdings dating back to the 19th century. Ranchers want to keep the working range and retain a way of life; environmentalists want to push for the wild, whether legally delineated or de facto; the federal agencies are converting to a doctrine of ecosystem management. The pieces in play are many, and seem irreconcilable, but the MBG attempts to keep the separate tiles whole while assembling them into a functioning mosaic. It's an exercise in environmental politics rather than philosophies.[1]

Proponents have begun to speak of a "working wilderness," an oxymoron only to ideologues. It means ranchers must accept that the old ways trashed landscapes and must incorporate modern science instead of folklore and embrace thinking about land ethics instead of reaching for hired guns. It means environmentalists must accept that people will not be swept from those lands and that conservation remedies may well involve active measures by them. Laissez-faire will not suffice for environmentalism anymore than for ranching. But however they engage the land, the players must engage one another. The upshot is less a melting pot than a rising magma chamber full of chunky half-ingested country rock. A working wilderness is to land management what a managed wildfire is to fire management.

What first joined them was fire on July 2, 1991, a burn that ranchers wanted to propagate—in fact, one had purposely though surreptitiously set it—but the Forest Service followed policy and law and suppressed it anyway. The local ranching community was outraged and set up the conversations that led to the awkward congregation of interests that became the MBG. And fire remains part of what holds them together—the prospects for good fire, the threat of bad fire, particularly the bad fire of fossil fuel-driven sprawl. Both need to engage open flame done right. It can't be done individually—too many edges, too much abrasion, too costly, too little variability to accommodate the diversity of biotas and flames.

The Malpai Borderlands Group is probably unique, and unreplicable in its particulars. Apart from pivotal people, a critical element was a vast holding at its center, the Grey Ranch, once a national forest, then traded into private hands, that encompassed Animas Mountain—a large core that could hold the pieces in its force field. When it went on the market, the Nature Conservancy bought it to keep it from being subdivided, and when federal purchase became politically impossible, sold it to the non-profit Animas Foundation established for the purpose. That defused the rancor over possible control by any one player.

But on a larger scale, that is what the Southwest requires generally. One on one its many pieces generate too much friction, heat rises, and the gears gag. Since the pieces won't dissolve and can't be removed, the best way to overcome that internal friction is to expand the domain of operation, lifting the pressure from piece-on-piece grating into a larger political and biophysical realm that allows each enclave its place yet permits collective action by the whole.

On a larger scale that is not a bad characterization of the American fire scene overall, with its chunks and regions that struggle to find common ground. What the Southwest brings to the table is a long history of parceling into pieces that resist change, and then slamming them together in ways that cannot be disguised by murky skies or thickened woods. Like the skyline of the Chiricahuas or the rim of the Grand Canyon, they appear without transitions.

In 2011 two spring fires showed contrasting visions of the future in southwestern wildlands. They displayed the two ways fire is being restored. Nor are they simply experiments; they are now part of an emerging historical record.

On April 28 a human-caused fire broke out in the Gila Wilderness. It was an area that had, in parts, for nearly 30 years, been prescribe burned and scorched by confined-and-contained wildfires. In places the Miller fire reburned sites for the fourth time since the fire revolution began—a percentage close to historic averages before livestock trampled the old fire regime. There were sporadic blowouts of course, but mostly the flames washed over the landscape like a monsoon shower. When southwesterly winds pushed the Miller fire toward the Gila Cliff Dwellings National Monument and clusters of forest campgrounds along Highway 15, it was fought. The Park Service had long prepared for such an event with fuels treatments. Engines protected structures, crews burned out along lines. It was a mixed response: this was not one integral fire but a fire of parts, each of which could merit a distinctive action.

Before it ended the Miller fire had scorched 88,835 acres. This was hardly a megafire by the evolving criteria of the new millennium, but it was big by most standards. The contrast between its story and that across the border on the Apache National Forest is more than striking: it's like a slap in the face. If the Miller fire spread like shadows through the woods, the Wallow fire blasted into the stratosphere like an eruption and threatened towns.

On June 2, 2011, the Wallow fire started from an untended campfire in the Bear Wallow Wilderness on the Apache National Forest, blew up, saturated the basin, and leaped over the rim. The next day it doubled in size, burning like a shot arrow to Alpine. By June 4 it was the largest single fire in recorded modern Arizona history. It burned the way all major wildfires in the region have. The long dry months before the summer monsoon readied the forests, the days lengthened and humidity shriveled, the prevailing winds from the southwest strengthened—the Wallow fire ran the same way, at the same times, as the Cerro Grande fire of 2000, the Rodeo-Chediski fire of 2002, the Warm fire of 2006. This is the time when fires get big and this is the direction they move. At the same time the Horseshoe II fire burned the Chiricahuas and the Monument fire the Huachucas. The only surprise is that fire officers thought the Wallow fire might stay in its spring den rather than rouse itself and, ravenous, go on a wild prowl for food. A map of burned area resembled nothing so much as a giant grizzly paw print raking over the landscape. The final tally included 538,000 acres.

The Miller fire came atop 30 years of steady restoration. The Wallow fire blasted over lands reclassified in purpose but not reworked to accommodate new fire regimes. In both the Mogollon and White mountains, fire is inevitable. But these fires, at these times, at these places, under these circumstances, were not. They reflected choices made about how to cope with fire, and about what restoration might mean. Did it mean returning to the presettlement era before Texas cattle and Hispanic sheep ate out the old order? Or did it mean adapting historic lessons to an

Anthropocene that promised a worsening climate, invasive grasses, a collision between a culture informed by internal combustion and one based on open fire? Did it mean a mash-up of the sort that drove ideologues and 10 a.m. veterans crazy, or was it a pragmatic way to cope with bad fire and encourage more good? It's worth pondering that only a scratch line in the duff separates travesty from tragedy.

All fire strategies fail. Prescribed fire can slip its leash, or fail to do the burning as desired. Natural fires can bolt out of control and burn in unnatural ways, and once awry can be hugely expensive and damaging to contain. Suppression cannot stop all fires, and where successful it can destabilize ecosystems. Landscaping does not eliminate fire, or even large fires, but changes its character, and if done poorly may destroy the values under protection. These are universal considerations, not unique to the Southwest. What makes the Southwest mosaic special is the antiquity of some of its tiles and the sharpness of their edges, even as new enclaves emerge and old borders break. Everything is visible. Nothing seems to dissolve away completely.

In recent years regional politics have been convulsed by a wave of immigration, much illegal, along with drug and human trafficking, that has washed across the Mexican border, broken down the old patterns, and threatened to overwhelm social arrangements. The environmental equivalent is a surge of fire that flares outside the inherited matrix and requires interventions of a particular sort. It won't be stopped by denunciation and suppression alone: the borders can't be hermetically sealed. The causes behind both trends are many, the choices for response compromised.

The paradox is that the Southwest has thrived on immigration. Growth powered by newcomers is, in fact, its primary industry, and one reason why Arizona's economy has outpaced New Mexico's. But the new wave is, it seems, too much and too different. So it is with fire. The swelling of burned area is not itself a problem: the land needs much more fire. The problem is that it is coming from too much bad fire and not enough good and without adequate say over who gains benefits and suffers losses.

In the postwar era the region, especially Arizona, has exerted an exaggerated influence on national politics. At one time the state supplied two sitting Supreme Court justices, two influential two-term secretaries of the interior, and a stream of presidential candidates. Something in the air, if not the water, seems to push regional concerns into the national scene. Local noise becomes national news. And so it continues to prove, not only with immigration, but with fire. It's not possible to hide or divert the flames any more than one could level the Mogollon Rim or erase chaparral from the Weaver Mountains.

Returning from the Tonto Rim one August afternoon, I watched a thunderhead collapsing over Mazatzal Peak. It began with grey veils of virga, then thickened into dark drapery as rain thundered down and smothered the mountain. Yet a slight twist of my head brought into view another summit further south, Four Peaks, bright with sun and sending streamers of smoke from a loose-herded wildfire burning through the regrowth of the Lone fire. One mountain range, two opposing processes, each distinct yet together making a dialectic of fire and water—a single panorama held them both. That I witnessed the scene from an SUV on a macadamized highway funded by a tax on gasoline completes the cycle of combustion.

It was classic Southwest spectacle. The scene displayed a clarity of contrast without inviting a similar clarity of comprehension. Each had its integrity, each had its sphere, both joined by long rhythms of time and sharp delineations of space. It's the kind of mash-up that will likely characterize American wildland fire in the years to come. It didn't happen first in the Southwest. It just happens here with unblinking undeniability. What to outsiders seem like contradictions, locals recognize as paradoxes. A region that more than any other preserves its past may also point, as few others can, to the future.

NOTE ON SOURCES

AS WITH THE OTHER VOLUMES of *To the Last Smoke*, my primary inspiration was my travels and conversations with local fire folk. To this cache of materials I have tried to introduce a historical backstory, and for this enterprise I needed more documentation than local file cabinets and memories could provide. Fortunately the region is well endowed with famous science, scholarship, and art.

The three books that most shaped my understanding come from geography, anthropology, and political history, respectively: Edward H. Spicer, *Cycles of Conquest: The Impact of Spain, Mexico, and the United States on the Indians of the Southwest, 1533–1960* (Tucson: University of Arizona Press, 1962); D. W. Meinig, *Southwest: Three Peoples in Geographical Change, 1600–1970* (New York: Oxford University Press, 1971); and Howard R. Lamar, *The Far Southwest, 1846–1912: A Territorial History* (Albuquerque: University of New Mexico Press, 2000). I also found Christopher J. Huggard and Arthur R. Gómez, eds., *Forests Under Fire: A Century of Ecosystem Mismanagement in the Southwest* (Tucson: University of Arizona Press, 2001) a useful survey of public lands, their controversies, and their institutional settings.

The scientific literature dates back to surveys by the Army Corps of Topographical Engineers. Government documents continue to furnish a substantial substrate for understanding. See, for example, Peter F. Ffolliott et al., eds., *Effects of Fire on Madrean Province Ecosystems: A Symposium Proceedings*, General Technical Report RM-GTR-289 (Fort

Collins, CO: USDA Forest Service, 1996), along with the serial conferences on the Madrean Archipelago listed in the notes to the prologue.

Very general, but helpful at times, was John L. Vankat, *Vegetation Dynamics on the Mountains and Plateaus of the American Southwest* (New York: Springer, 2013). Peter Friederici, ed., *Ecological Restoration of Southwestern Ponderosa Pine Forests* (Washington, DC: Island Press, 2003) is an indispensable encyclopedia for its topic.

The most surprising lapse (for me) was literature. The Southwest has a long and fabled literary heritage, with numerous compendia, but I found the surveys less helpful than I hoped. I was better served by starting with the sources.

For more particular references consult the notes attached to each essay.

NOTES

PROLOGUE

1. The Southwest has attracted many historians, geographers, anthropologists, literary critics—nearly every branch of scholarship, it seems. Some of the works that most influenced my own thinking are D. W. Meinig, *Southwest: Three Peoples in Geographical Change 1600–1970* (New York: Oxford University Press, 1971) and Edward H. Spicer, *Cycles of Conquest: The Impact of Spain, Mexico, and the United States on the Indians of the Southwest, 1533–1960* (Tucson: University of Arizona Press, 1962), from whose title derives my own. On state histories I relied on Thomas Sheridan, *Arizona: A History*, rev. ed. (Tucson: University of Arizona Press, 2012); Marc Simmons, *New Mexico: An Interpretive History* (Albuquerque: University of New Mexico Press, 1988); and Howard R. Lamar, *The Far Southwest, 1846–1912: A Territorial History* (New York: Norton, 1966). For the Borderlands overall, I consulted David J. Weber, *The Spanish Frontier in North America: The Brief Edition* (New Haven, CT: Yale University Press, 2009).
2. The best syntheses of regional biogeography (excluding the Colorado Plateau) are three conference proceedings: Leonard DeBano et al., *Biodiversity and Management of the Madrean Archipelago: The Sky Islands of Southwestern United States and Northwestern Mexico*, General Technical Report RM-GTR-264, U.S. Forest Service, 1995; Gerald Gottfried et al., *Connecting Mountain Islands and Desert Seas: Biodiversity and Management of the Madrean Archipelago II*, Proceedings RMRS-P-36, U.S. Forest Service, 2004; and Gerald Gottfried et al., *Merging Science and Management in a Rapidly Changing World: Biodiversity and Management of he Madrean Archipelago III and 7th Conference on Research*

and Resource Management in the Southwestern Deserts, Proceedings RMRS-P-67, U.S. Forest Service, 2013.

3. For a general background of southwestern fire, see Stephen J. Pyne, *Fire in America: A Cultural History of Wildland and Rural Fire* (Seattle: University of Washington Press, 1997), 514–29, and "Nouvelle Southwest," in *World Fire* (Seattle: University of Washington Press, 1997), 282–95; and John Herron, "'Where There's Smoke': Wildfire Policy and Suppression in the American Southwest," in *Forests under Fire: A Century of Ecosystem Mismanagement in the Southwest*, ed. Christopher J. Huggard and Arthur R. Gómez (Tucson: University of Arizona Press, 2001), 181–210. Holsinger quote (and other useful passages) from "The Boundary Line Between Desert and Forest," *Forestry and Irrigation* 8 (1902): 23–25.
4. Pyne, *World Fire*, 283.
5. The Southwest is singularly blessed by the presence of the Laboratory of Tree-Ring Research, a world-class facility that has perfected its techniques on the region's forests, and particularly its fires. The lab's fire-scar histories are a foundational source for any study of regional fire history.
6. On the Forest Service presence, see Edwin A. Tucker and George Fitzpatrick, *Men Who Matched the Mountains: The Forest Service in the Southwest* (Washington, DC: Government Printing Office, 1972) and Robert D. Baker et al., *Timeless Heritage: A History of the Forest Service in the Southwest*, FS-409, U.S. Forest Service, 1988.
7. The ponderosa pine forests of the Southwest helped underwrite much of the forest health crisis the boiled over in the early 1990s. A useful overview for the Colorado Plateau generally is Gary Paul Nabhan, Marcelle Coder, and Susan J. Smith, *Woodlands in Crisis: A Legacy of Lost Biodiversity on the Colorado Plateau*, Bilby Research Center Occasional Papers No. 2, Bilby Research Center, 2004. Regarding the pines in particular see the various conference proceedings on the topic, for example, Merrill R. Kaufmann et al., *Old-Growth Forests in the Southwest and Rocky Mountain Regions: Proceedings of a Workshop*, General Technical Report RM-213, U.S. Forest Service, 1992.

THE JEMEZ

1. A field trip sponsored by the Joint Fire Science Program for its board on October 29, 2014, provided a marvelous setting from which to think about fire in the Jemez Mountains. I had the privilege of conversing further with Bill Armstrong of the Santa Fe National Forest and the

next day with Craig Allen of the U.S. Geological Survey, who graciously shared his deep knowledge of the region.

2. The best introduction to fire in the Jemez is Craig D. Allen's "Lots of Lightning and Plenty of People: An Ecological History of Fire in the Upland Southwest," in *Fire, Native Peoples, and the Natural Landscape*, ed. Thomas R. Vale (Washington, DC: Island Press, 2002), 143–93.
3. Teralene S. Foxx, *La Mesa Fire Symposium*, LA-9236-NERP, UC-11, Los Alamos National Laboratory, February 1984; Craig D. Allen, *Fire Effects in Southwestern Forests: Proceedings of the Second La Mesa Fire Symposium*, General Technical Report RM-GTR-286, U.S. Forest Service, 1996.

THE MOGOLLONS

1. No national forest on my travels has rivaled the Gila as a host. For this I wish to thank Andrea Martinez who set up an astonishing meeting with Steve Servis, Gary Benavidez, Albert Holguin, Mike Noel, Toby Richards, and Ellen Brown. Paul Boucher and Gabe Holguin contributed with subsequent phone interviews. Forest Supervisor Kelley Russell graciously added her thoughts. Brian Park worked up the forest's fire statistics. My notes quickly sprawled beyond their binders. I'm grateful to all.
2. Thomas W. Swetnam and John H. Dieterich, "Fire History of Ponderosa Pine Forests in the Gila Wilderness, New Mexico," in *Proceedings—Symposium and Workshop on Wilderness Fire*, GTR-INT 182, U.S. Forest Service, 1985, 390–97.
3. Aldo Leopold, "Grass, Brush, Timber, and Fire in Southern Arizona," *Journal of Forestry* 22 (October 1924): 1–10; see also "Virgin Southwest," in *The River of the Mother of God and Other Essays by Aldo Leopold*, ed. Susan L. Flader and J. Baird Callicott (Madison: University of Wisconsin Press, 1991), 179. Circular quote from Leopold, "To the Forest Officers of the Carson," 45, and Sapello quote from "Conservation in the Southwest," 92, in Flader and Callicott, *River of the Mother of God*. Leopold comments on prescribed burning from "'Piute Forestry' vs. Forest Fire Protection," in *Aldo Leopold's Southwest*, ed. David E. Brown and Neil B. Carmony (Albuquerque: University of New Mexico Press, 1995), 139–46.
4. Mullin quoted in Robert Baker et al., *Timeless Heritage: A History of the Forest Service in the Southwest*, FS-409, U.S. Forest Service, 1988, 161.

Also useful for the early days is Edwin A. Tucker and George Fitzpatrick, *Men Who Matched the Mountains: The Forest Service in the Southwest*, U.S. Forest Service, Southwestern Region, 1972. For a portrait during the heyday of suppression, see Randle Hurst, *The Smokejumpers* (Caldwell, ID: Caxton Printers, 1966).

5. Gila NF, *Forest Multiple Use Guide* (1975), 3–45.
6. Adam Burke, "Keepers of the Flame," *High Country News*, November 8, 2004, https://www.hcn.org/issues/286/15102.
7. For a sketch of what happened, see Don R. Webb and R. L. Henderson, "Gila Wilderness Prescribed Natural Fire Program," in *Proceedings—Symposium and Workshop of Wilderness Fire*, ed. J. E. Lotan et al., General Technical Report GTR-INT-182, U.S. Forest Service, 1985, 413–14.
8. Information mostly by conversation with Steve Servis, but see Stephen H. Servis and Janet F. Hurley, "Appropriate Suppression Response on the Gila National Forest," in *Effects of Fire in Management of Southwestern Natural Resources*, ed. J. S. Krammes, General Technical Report RM-GTR-191, U.S. Forest Service, 1990, 244–45.
9. Steve Servis and Paul F. Boucher, "Restoring Fire to Southwestern Ecosystems: Is It Worth It?," in *Symposium on Fire Economics, Planning, and Policy: Bottom Lines*, General Technical Report PSW-GTR-173, U.S. Forest Service, 1999, 247–53.
10. Several published articles speak to the Gila story: Josh McDaniel, "The Fire Laboratory: Forest Restoration on the Gila," Southwest Fire Science Consortium; Molly Hunter, "Wildland Fire Use in Southwestern Forests: An Underutilized Management Option?," *Natural Resources Journal* 47, no. 2 (2007): 257–66; Matt Rollins et al., "Final Report; Historical Wildland Fire Use: Lessons to be Learned from Twenty-Five Years of Wilderness fire Management," Joint Fire Science Program Project Number 01-1-1-06 (October 31, 2007); and Molly E. Hunter, Leigh B. Lentile, and Jose M. Iniguez, "Monitoring Effectiveness of Prescribed Fire and Wildland Fire Use in the Gila National Forest, New Mexico," Joint Fire Science Program Project Number 08-1-1-10, n.d.
11. Boucher quoted in Burke, "Keepers of the Flame."
12. Bill Derr and Roger Seewald, "Congressional Review of Several Fires in New Mexico During May and Early June, 2012: Seewald Report" (2013), and William A. Derr, "Wildfire Review Report: Whitewater-Baldy Complex and Little Bear Fires: New Mexico—May thru June 2012." There were other spontaneous reports from former fire-

management officers or residents of the area, all lamenting the losses of what they had known in their youth or over their careers; see, for example, Allen Campbell, "Whitewater Fire, a Lasting Legacy," at http://pearce.house.gov/sites/pearce.house.gov/files/WHITEWATER%2C_the_legacy_PDF.pdf. The counterview was these fires were destined to happen and rebirthed a dying forest.

13. George R. Stewart, *Fire* (New York: Random House, 1948).

THE HUACHUCAS

1. For an example of the consensus (and unlikely alliance) that resulted, see Craig D. Allen et al., "Ecological Restoration of Southwestern Ponderosa Pine Ecosystems: A Broad Perspective," *Ecological Applications* 12, no. 5 (2002): 1418–33.
2. Shelley R. Danzer, Chris H. Baisan, and Thomas W. Swetnam, "The Influence of Fire and Land-Use History on Stand Dynamics in the Huachuca Mountains of Southeastern Arizona," in *Effects of Fire on Madrean Province Ecosystems: A Symposium Proceedings*, ed. Peter F. Ffolliott et al., General Technical Report RM-GTR-289, U.S. Forest Service, 1996, 265–70.

THE KAIBAB

1. For general standards, see "Wildland Fire Use Implementation Procedures Reference Guide (May 2005), signed off by all the federal land agencies. For the local scene, see "Kaibab National Forest Fire Management Plan" (October 2005), which includes the specifics of WFUs.
2. Prescribed natural fire review submitted to the Fish and Wildlife Service (1999), p. 3; "no significant impact" quote from "Kaibab National Forest Land Management Plan, as amended." Amendment 4 (August 28, 2000) applies to the Wildland Fire Use Plan environmental assessment.
3. Aldo Leopold, *A Sand County Almanac* (Oxford: Oxford University Press, 1949), 130–32; and Christian Young, *In the Absence of Predators: Conservation and Controversy on the Kaibab Plateau* (Lincoln: University of Nebraska Press, 2002), 201.
4. An excellent summary of burned history since 1959 is available in Garrett W. Meigs, "Recent Patterns of Large Fire Events on Kaibab Plateau, Arizona, USA" (honors thesis, Cornell University, Department of Natural Resources, May 2004). I rely on Meigs for my statistical analysis.

On Covington's park research, see Peter Z. Fulé et al., "Assessing Fire Regimes on Grand Canyon Landscapes with Fire-Scar and Fire-Record Data," *International Journal of Wildland Fire* 12 (2003): 129–45, and Fulé et al., "Comparing Ecological Restoration Alternatives: Grand Canyon, Arizona," *Forest Ecology and Management* 170 (2002): 19–41, both of which provide general bibliographies.

5. Quote from "Warm Fire After Action Review, North Kaibab Ranger District, Kaibab National Forest, October 11, 2006," 1.
6. USDA Office of Inspector General, Western Region, "Audit Report: Forest Service Large Fire Suppression Costs," Report No. 08601-44-SF, November 2006, 3.
7. W. W. Covington, notes from conversation with Regional Forester Forsgren, 11:30 p.m., Monday, February 12, 2007.
8. "WHAT WENT WRONG??? Warm Wildland Fire Use—Wildfire," draft July 9, 2006, North Kaibab Ranger District, Kaibab NF.

RHYMES WITH CHIRICAHUA

1. The summa remains Paul S. Martin and Richard Klein, eds., *Quaternary Extinctions: A Prehistoric Revolution* (Tucson: University of Arizona Press, 1984). An excellent distillation applied to the region is Paul S. Martin, "Ghostly Grazers and Sky Islands," in *Connecting Mountain Islands and Desert Seas: Biodiversity and Management of the Madrean Archipelago II*, comp. Gerald J. Gottfried et al., RMRS-P-36, U.S. Forest Service, 2005, 26–34.
2. Sources: Henry F. Dobyns, *From Fire to Flood: Historic Human Destruction of Sonoran Desert Riverine Oases* (Socorro, NM: Ballena Press, 1981); and, though dating, still useful as a thumbnail, Stephen Pyne, *Fire in America* (Seattle: University of Washington Press, 1997), 514–29.
3. Dobyns, *From Fire to Flood*, 36.
4. Edward H. Spicer, *Cycles of Conquest: The Impact of Spain, Mexico, and the United States on the Indians of the Southwest, 1533–1960*, 2nd ed. (Tucson: University of Arizona Press, 1967).
5. Dobyns, *From Fire to Flood*, 30.
6. In addition to *The Changing Mile*, see two other classic studies: Robert R. Humphrey, *90 Years and 535 Miles: Vegetation Changes Along the Mexican Border* (Albuquerque: University of New Mexico Press, 1987), and Conrad Joseph Bahre, *A Legacy of Change: Historic Human Impact*

on Vegetation in the Arizona Borderlands (Tucson: University of Arizona Press, 1991).

TOP-DOWN ECOLOGY

1. This essay was part of my early experiments in fire journalism, first written in 2009. Much of what it contains is not available in the published literature, but was acquired by a marvelous field tutorial organized by Peter A. Gordon, fire officer for the Coronado National Forest. I wish to thank Pete, Chris Stetson, Buddy Zale, Toni Strauss, and especially the deeply knowledgeable and quietly passionate Randall Smith for their time and willingness to share their experiences. Any errors of fact or interpretation are of course mine alone.
2. For an excellent summary of the events, see Paul Hirt, "Biopolitics: A Case Study of Political Influence on Forest Management Decisions, Coronado National Forest, Arizona," in *Forests Under Fire: A Century of Ecosystem Mismanagement in the Southwest*, ed. Christopher J. Huggard and Arthur R. Gómez (Tucson: University of Arizona Press, 2001). For a collection of essays on the topic, see Conrad A. Istock and Robert S. Hoffman, eds., *Storm over a Mountain Island: Conservation Biology and the Mt. Graham Affair* (Tucson: University of Arizona Press, 1995).
3. Peter Frost, "Hot Topics and Burning Issues: Fire as a Driver of System Processes—Past, Present, Future" (paper presented at a postgraduate course offered by the C. T. de Wit Graduate School for Production Ecology and Resource Conservation [PE&RC] at Wageningen University, the Global Fire Monitoring Center/Max Planck Institute for Chemistry, and the United Nations University, March 30–April 5, 2008), 2.

THE VIEW FROM TANQUE VERDE

1. Thanks to Perry Grissom for an informative chat and a generous cache of historic documents. The park is unusual both in having records and in making them available; I'm grateful for both. Thanks, too, to the National Advanced Fire and Resources Institute's Fire in Ecosystem Management (M-580) course, which hosted a field trip to Saguaro and the Santa Catalinas. Of course no one I talked to is responsible for my opinions or interpretation.

2. Basic information from *Saguaro National Park Fire Management Plan* (July 2007); "Saguaro National Park, Administrative Timeline Relating to Fire"; National Park Service, Fire Management Program, background paper; Saguaro, *Historic Resource Study*, Chapter 4 (F), Forest Fire Policy; Pamela J. Swantek et al., "GIS Database Development to Analyze Fire History in Southern Arizona and Beyond: An Example from Saguaro National Park," *Technical Report No. 61*, U.S. Geological Survey, Cooperative Parks Study Program, University of Arizona, 1999.
3. See Louis L. Gunzel, "National Policy Change...Natural Prescribed Fire," *Fire Management* 35, no. 3 (Summer 1974): 6–8, and Dave Gibson, "Saguaro National Monument Prescribed Fire Plan" (honors thesis, BSc Forestry, University of Arizona, 1974).
4. Gunzel, "National Policy Change," 8.
5. Figures from Saguaro National Park, "Briefing Statement: Fire Management Program," May 28, 2004.

REINVENTING A FIRE COMMONS

1. Thanks to Henry Provencio, Mary Lata, and Neil McCusker for an engaging discussion of 4FRI, and to Earl Stewart for comments on a draft. I have framed the program in terms different from theirs, and I doubt they will agree with all I say. But, as I believe 4FRI demonstrates, we don't need unanimity on principles or antecedents. We need only agree on their outcomes and the means to achieve them.
2. For background on the research that underpinned 4FRI, Pyne, "Doc Smith's History Lesson," in *Smokechasing* (Tucson: University of Arizona Press, 2003), and a profile of its founder in "The Kaibab: Friendly Fire," in this volume. Also useful as a summary of the early science behind the Flagstaff model—and the evolution of a social mechanism to propagate it—is Peter Friederici, ed., *Ecological Restoration of Southwestern Ponderosa Pine Forests* (Washington, DC: Island Press, 2003).
3. The 4FRI website gives a thorough overview of the history and documentation behind the program. See http://www.fs.usda.gov/4fri. That introduction extends to the CFLRP program that authorized and funds it, and links to the *Statewide Strategy for Restoring Arizona's Forests*.
4. For the uninitiated "dbh" stands for "diameter at breast height," which is standardized at 4.5 feet.

5. U.S. Forest Service, "Draft Record of Decision for the Four-Forest Restoration Initiative, Coconino and Kaibab National Forests, Coconino County, Arizona" (November 2014).
6. I have relied primarily on Claudine LoMonaco, "Lost in the Woods: How the Forest Service is Botching its Biggest Restoration Project," *High Country News*, September 1, 2014, 12–20. My attempts to discuss the matter with 4FRI participants have gone nowhere. They want, naturally enough, to discuss the new contract and its promise for the future.

THINKING LIKE A BURNT MOUNTAIN

1. For the background on how the kill site was identified, see Susan Flader, "Searching for Aldo Leopold's Green Fire," *Forest History Today*, Fall 2012, 26–34.

UNDER THE TONTO RIM

1. The Tonto staff went out of its way to inform me. Bill Hart helped set up a field trip, and then conversed for several hours about the forest's efforts (and, at my prompting, his own career in fire). On the Payson District I had the benefit of on-site conversations with Angela Elam, the district ranger; Don Nunley, the FMO; and Jason Cress, the fuels assistant fire management officer. They were knowledgeable and helpful, and any breakdowns in communicating what I learned are my doing. I thank them all.
2. On the feuds, see Don Dedera, *A Little War of Our Own: The Pleasant Valley Feud Revisited* (Flagstaff, AZ: Northland Press, 1988), and for broader background, Tomas E. Sheridan, *Arizona: A History*, rev. ed. (Tucson: University of Arizona Press, 2012), and Aldo Leopold, "Grass, Brush, Timber, and Fire in Southern Arizona," in Flader and Callicott, *River of the Mother of God*, 121.
3. Figures come from Candace C. Kant, *Zane Grey's Arizona* (Flagstaff, AZ: Northland Press, 1984), 5, 137.
4. From Kant, *Zane Grey's Arizona*, 37.
5. Adam Sowards gives an excellent overview in "Range Rivalries: An Environmental and Cultural History of Arizona's Tonto National Forest Region" (master's thesis, Arizona State University, May 1997); see p. 70 for the photo of the cow; Coxen also quoted on p. 70. A distilled

overview of range management is available in Eddie Alford, "Tonto Rangelands—A Journey of Change," *Rangelands* 15, no. 6 (December 1993): 261–68.

Grey's jeremiad was published in *Outdoor America* (November 1924). For the presettlement fire history, see Mark Kaib, "Fire History Reconstructions in the Mogollon Province Ponderosa Pine Forests of the Tonto National Forest, Central Arizona," final report, October 2001, Laboratory of Tree-Ring Research, University of Arizona, on file with Tonto National Forest. Leopold quote from "The Virgin Southwest," in Flader and Callicott, *River of the Mother of God*, 179.

6. Gifford Pinchot, *Breaking New Ground* (Seattle: University of Washington Press, 1972), 179.
7. See the website for the Dude fire staff ride for a library of documents, maps, and videos at http://www.fireleadership.gov/toolbox/staffride/library_staff_ride11.html. Particularly useful is Michael A. Johns, "The Dude Fire" (2009), and "Accident Investigation Report: Dude Fire Incident, Multiple Firefighter Fatality, June 26, 1990, Southwest Region, Tonto National Forest."
8. Figures from Tonto National Forest, Payson Ranger District, Fuels Program Brief (August 4, 2014).

SQUARING THE TRIANGLE

1. Acknowledgements must begin with Bil Grauel, who helped set up a visit to San Carlos. Then, so many at San Carlos to thank: Dan Pitterle, Dee Randall, Clark Richins, Kelly Hetzler, Bob Hetzler, Seth Pilsk, and Duane Chapman, all of whom took time out of their busy workday to introduce me to how they see fire on their part of the Earth and in doing so, made Point of Pines a reference point for what the future of fire in the American West might be.
2. Useful background information is available in Dan Pitterle, ed., "San Carlos Apache Indian Reservation Wildland Fire Management Plan." *Programmatic Environmental Assessment*. EA No. FO-SCA-EA-02-02 (January 2003), and Dan Pitterle, ed., "San Carlos Apache Indian Reservation Wildland Fire Management Plan" (January 2003).
3. Observations on western Apache fire practices from conversation with Seth Pilsk.
4. The best summary of fire from tree rings in recent centuries is Mark Kaib, "Fire History in Mogollon Province Ponderosa Pine Forests of

the San Carlos Apache Tribe, Central Arizona" (master's thesis, University of Arizona, 2001). For contemporary developments, see Kim Kelly et al., "Restoring and Maintaining Resilient Landscapes Through Planning, Education, Support, and Cooperation on the San Carlos Apache Reservation: A Historical, Cultural, and Current View" (May 8, 2013).

5. Sheridan, *Arizona*, gives the barebones chronology. For background ethnography, see Richard J. Perry, *Western Apache Heritage: People of the Mountain Corridor* (Austin: University of Texas Press, 1991). A famous inquiry into Western Apache relations with their land is Keith Basso, *Wisdom Sits in Places: Landscape and Language Among the Western Apache* (Albuquerque: University of New Mexico Press, 1996), which raises interesting questions about how these peoples built, and rebuilt, their cultural connection to the lands they were given.
6. For a synopsis of grazing history, see Harry T. Getty, "Development of the San Carlos Apache Cattle Industry," *The Kiva* 3, no. 3 (February 1958): 1–4; and Harry T. Getty, *The San Carlos Indian Cattle Industry*, Anthropological Papers of the University of Arizona, no. 7 (Tucson: University of Arizona Press, 1963).
7. On the timber industry, see Jack August, Jr., Art Gomez, and Elmo Richardson, "From Horseback to Helicopter: A History of Forest Management on the San Carlos Apache Reservation" (American Indian Resource Organization, 1984).
8. Harold Weaver, "Fire as an Ecological Factor in the Southwestern Ponderosa Pine Forests," *Journal of Forestry* (February 1951): 93–98, quote from 95.
9. Information from Bob Gray, emails to author dated August 1, 2014, and March 12, 2016. The heritage of this transitional era seems to have been ignored or forgotten by the cohort that since established the modern era.
10. Harold Biswell et al., *Ponderosa Fire Management: A Task Force Evaluation of Controlled Burning in Ponderosa Pine Forests of Central Arizona* (Tallahassee, FL: Tall Timbers Research Station, 1973). For a brief chronicle of significant events in fire management, see also Kelly et al., "Restoring and Maintaining."
11. See Pitterle, "San Carlos."
12. Quotes and numbers from William Grauel, e-mail to author on September 11, 2013.
13. Emil W. Haury, *Point of Pines, Arizona: A History of the University of Arizona Archaeological Field School*, Anthropological Papers of the University of Arizona, no. 50 (Tucson: University of Arizona Press, 1989), 15.

14. Haury, *Point of Pines*, 15.
15. Haury, *Point of Pines*, 58, 63.
16. J. F. Lasley, J. T. Montgomery, and F. F. McKenzie, "Artificial Insemination in Range Cattle: A Preliminary Report," *Journal of Animal Science* (1940): 102–5 for the technical origins.

    On the square kiva, see Haury, *Point of Pines*, 46.
17. William Grauel, "Long Term Weather, Fire Behavior, and Risk Assessment, Skunk Fire, San Carlos Agency, June 5, 2014" gives the most useful chronology and description of conditions. The basics were supplemented by discussions with Dan Pitterle, Bob Hetzler, and Duane Chapman.
18. Statistics on the Skunk fire from Bureau of Indian Affairs, San Carlos Agency, "Skunk Fire, Day 48, Incident Action Plan, Thurs, June 5, 2014."

A REFUSAL TO MOURN THE DEATH, BY FIRE, OF A CREW IN YARNELL

1. In an eerie echo recall how Ahab is first described: a burned, lightning-struck snag, with a "slender rod-like mark, lividly whitish" running down a body that "looked like a man cut away from the stake, when the fire has overrunningly wasted all the limbs without consuming them."
2. Quotes from Wildland Fire Lessons Learned Center, *Two More Chains*, vol. 3, no. 2 (Summer 2013): 2.

EPILOGUE

1. The MBG story has been told many times to many purposes. An excellent survey is Nathan F. Sayre, *Working Wilderness: The Malpai Borderlands Group and the Future of the Western Range* (Tucson, AZ: Rio Nuevo Publishers, 2005). On FireScape see the program website: http://www.azfirescape.org.

INDEX

ABOUT THE AUTHOR

Stephen J. Pyne is a historian in the School of Life Sciences, Arizona State University. He is the author of over 20 books, mostly on wildland fire and its history but also dealing with the history of places and exploration, including *The Ice*, *How the Canyon Became Grand*, and *Voyager*. His current effort is directed at a multivolume survey of the American fire scene—*Between Two Fires: A Fire History of Contemporary America* and *To the Last Smoke*, a suite of regional reconnaissances, all published by the University of Arizona Press.